To my father, who taught me to write

Stephanie Frank Singer

Symmetry in Mechanics

A Gentle, Modern Introduction

Birkhäuser
Boston • Basel • Berlin

Stephanie Frank Singer
Philadelphia, PA
www.symmetrysinger.com

Library of Congress Cataloging-in-Publication Data

Singer, Stephanie Frank, 1964-
    Symmetry in mechanics : a gentle, modern introduction / Stephanie Frank Singer.
        p. cm.
    Includes bibliographical references and index.
    ISBN 0-8176-4145-9 (acid-free paper) – ISBN 3-7643-4145-9 (acid-free paper)
        1. Mechanics, Analytic. 2. Geometry, Differential. I. Title.

QA805.S62 2001
531–dc21
                                       00-049398
                                          CIP

AMS Subject Classifications: 70F05, 70H33, 70H05, 22E, 53C15, 53A25, 53C80

ISBN 0-8176-4145-9   Printed on acid-free paper.

©2001 Stephanie Frank Singer   **Birkhäuser**
©2004 Stephanie Frank Singer, 2nd printing

Printed in the United States of America.   (TXQ/Ham)

9 8 7 6 5 4 3 2   SPIN 11013433

Birkhäuser is a part of *Springer Science+Business Media*

www.birkhauser.com

Contents

Preface

"And what is the use," thought Alice, "of a book without pictures
or conversations in it?"

—Lewis Carroll

This book is written for modern undergraduate students — not the ideal students that mathematics professors wish for (and who occasionally grace our campuses), but the students like many the author has taught: talented but appreciating review and reinforcement of past course work; willing to work hard, but demanding context and motivation for the mathematics they are learning. To suit this audience, the author eschews density of topics and efficiency of presentation in favor of a gentler tone, a coherent story, digressions on mathematicians, physicists and their notations, simple examples worked out in detail, and reinforcement of the basics.

Dense and efficient texts play a crucial role in the education of budding (and budded) mathematicians and physicists. This book does not presume to improve on the classics in that genre. Rather, it aims to provide those classics with a large new generation of appreciative readers.

This text introduces some basic constructs of modern symplectic geometry in the context of an old celestial mechanics problem, the *two-body problem*. We present the derivation of Kepler's laws of planetary motion from Newton's laws of gravitation, first in the style of an undergraduate physics course, and then again in the language of symplectic geometry. No previous exposure to symplectic geometry is required: we introduce and illustrate all necessary constructs.

Sir Isaac Newton analyzed completely the two-body problem (and many other problems) in his famous and ultimate book, *Philosophiae Naturalis Principia Mathematica* [N]. From his own fundamental laws of motion and Kepler's laws summarizing observations of the planets, Newton deduced the inverse square law for the attractive gravitational force of the sun. Our analysis is more modest: we use the fundamental laws of motion and the inverse square law to deduce Kepler's laws. In the process Newton performs an early (the first, to the author's knowledge) symplectic reduction. He uses the concept of the center-of-mass and conservation of linear momentum to show that "the motions of bodies included in a given space are the same among themselves, whether that space is at rest, or moves uniformly forward in a right line without any circular motion" [N, Corollary V of "Axioms, or Laws of Motion"]. And while Newton's (and others', as found in most current physics textbooks) exploitation of the conservation of angular momentum is not, strictly speaking, symplectic reduction, the same results can be obtained by symplectic reduction. So the two-body problem provides two natural examples of symplectic reduction.

Chapter 0 covers some preliminary material. Chapter 1 presents the derivation of Kepler's laws of planetary motion from Newton's laws of gravitation in the style of a typical American undergraduate physics text. Chapter 8 presents the same argument in the language of modern symplectic geometry. The chapters in between develop the concepts and terminology necessary for the final chapter, providing a detailed translation between the quite different languages of mathematics and physics.

Warning

This book is not intended as a comprehensive introduction to symplectic geometry or classical mechanics. For strong undergraduate students it will not suffice as the sole text for a full semester course. Instead of formulating a general theory and treating a variety of problems, this text makes explicit the ties between mathematics and physics, as well as the ties between powerfully abstract formulations and concrete calculations. Like the trunk of a Japanese maple, it should support and unite readers who will branch off in various directions.

There are many excellent treatments of the two-body problem in the physics literature, and excellent treatments of symplectic reduction in the graduate mathematics literature. Readers desiring broader or more sophisticated texts should consult the Recommended Reading section. The book in your hands, accessible to anyone who has studied multivariable calculus and the rudiments of linear algebra, should provide a useful complement to the existing sources.

Guide to the Instructor

This book can be used as a supplement to courses on differential geometry or Lie theory. In particular, Chapters 3, 5 and 6 are self-contained units, with concrete examples of sophisticated ideas and useful exercises.

The book could be a major component of a course on symplectic geometry or classical mechanics. The book (in its entirety or with Chapter 3 omitted) could serve to structure the first part of a course in symplectic geometry, providing motivation for a more standard exposition of the mathematics. It would also be appropriate at the end of an example-driven semester course on classical mechanics, in which case students should be encouraged to work out the symplectic versions of examples treated earlier.

For more advanced courses in symplectic geometry, Chapters 1 and 8 link the mathematics to its historical roots and current physics terminology. Likewise, in an advanced physics course on mechanics these chapters would make the relevant mathematics more accessible.

Guide to the Reader

Readers who have no particular background in symplectic geometry or classical mechanics may wish to start at the beginning and read through, doing the exercises as they go. They should feel free to ignore anything marked "optional," including all of Chapter 3. They should not be discouraged by the occasional exercise requiring a branch of mathematics they may not know (such as techniques for solving differential equations), or by the occasional paragraph aimed at more sophisticated readers (usually discussing links to other areas of mathematics).

Readers working through a standard mathematical exposition of symplectic geometry may find this book a useful auxiliary source of explicitly calculated examples.

Readers familiar with mechanics or with symplectic geometry may wish to start with Chapter 8 to get the whole story quickly. The references in that chapter should make it easy to dip back into the rest of the text to fill any gaps in understanding.

The reader will find references to many other texts in the Bibliography and Recommended Reading section. Listed there are good books for solidifying the prerequisites, interesting collateral reading, alternative sources for the material presented, major classics in the field and research articles by currently active mathematicians.

Acknowledgments

Thanks to several people for specific contributions to this book: Karen Uhlenbeck and Chuu-Lian Terng for the invitation to give a series of lectures on symplectic geometry for undergraduates at the IAS/Park City Mentoring Program for Women in 1997 at the Institute for Advanced Study (IAS) in Princeton; Haverford College and the American Mathematical Society Centennial Fellowship program for financial support; Kathleen McGoldrick, Anne Humes, Catherine Jordan and the IAS staff for nontechnical support; Jerrold Marsden and Ann Kostant for their encouragement; Eugene Lerman for suggesting references; J. J. Duistermaat for copious historical insights; Ben Allen for writing solutions to exercises; and those who critiqued the manuscript along the way, especially Allen Knutson, Tanya Schmah, Alan Weinstein and the anonymous reviewers.

Thanks also to all those who taught me, supported me and believed in me, especially the community of symplectic geometers and my students at Haverford College.

Symmetry in Mechanics

A Gentle, Modern Introduction

0

Preliminaries

0.1 Notation and Conventions

Throughout the text the symbol ":=" indicates a defining equality. For example, to define the symbol $\mathbb{R}^3$ we write

$$\mathbb{R}^3 := \left\{ \begin{pmatrix} x \\ y \\ z \end{pmatrix} : x \in \mathbb{R},\ y \in \mathbb{R} \text{ and } z \in \mathbb{R} \right\}.$$

The symbol $\mathbb{R}^+$ denotes the set of strictly positive real numbers. That is,

$$\mathbb{R}^+ := \{ r \in \mathbb{R} : r > 0 \}.$$

We use boldface for row or column vectors such as $\mathbf{r}$ and $\mathbf{p}$. The letters x, y and z are the coordinate axes in $\mathbb{R}^3$. We almost always indicate components of three-vectors with subscripts x, y and z and only rarely with numbers. So a typical element of $\mathbb{R}^3$ is

$$\mathbf{r} = \begin{pmatrix} r_x \\ r_y \\ r_z \end{pmatrix},$$

while $\mathbf{r}_1$ and $\mathbf{r}_2$ denote two different three-vectors. In occasional discussions of $\mathbb{R}^n$ for an arbitrary natural number n we write $r = (r_1, \ldots, r_n)^T$. The T

superscript denotes the transpose. Italics indicate a word whose definition is nearby.

We use the standard notation of professional mathematicians for function definitions. Namely, we indicate the function name f, the domain D, the target space S as well as the rule for finding the output value $f(x)$ from any input value x in the domain via

$$f : D \to S$$
$$x \mapsto f(x).$$

For example, to describe the function g that squares any real number, we write

$$g : \mathbb{R} \to \mathbb{R}$$
$$x \mapsto x^2.$$

Note that the target space (in this case $\mathbb{R}$) is not necessarily equal to the range (in this case $\mathbb{R}^+ \cup \{0\}$). However, the target space must contain the range.

Finally, the name of the Norwegian mathematician who first used groups of transformations to study differential equations is Sophus Lie, pronounced "Lee."

0.2   Physics and Math Background

There are several ideas and formulas from multivariable calculus, linear algebra and basic physics that the reader will need. People who have completed the first two years of the standard college curriculum in mathematics and a year of college physics will have seen most of these. Readers can brush up and fill gaps with any standard textbooks or with our suggestions: Marsden, Tromba and Weinstein's *Basic Multivariable Calculus* [MTW], Lay's *Linear Algebra and Its Applications* [L], Strang's *Introduction to Linear Algebra* [St] and *The Feynman Lectures on Physics, Volumes I and II* [Fe].

From physics, the reader should be familiar with momentum, velocity and Newton's second law, force $=$ mass $\times$ acceleration, in three-dimensional space. Newton's law of gravitation plays a fundamental role. The reader should be comfortable with center-of-mass coordinates. For Section 4.1 the reader should understand kinetic and potential energy. In one inessential but interesting example (see Section 4.3) the expression $q(\mathbf{v} \times \mathbf{B})$ for the magnetic force produced by a magnetic field vector $\mathbf{B}$ on a particle of charge q moving with velocity $\mathbf{v}$ is useful.

The reader should be familiar with the basic notions of vector calculus, including matrix multiplication, the chain rule, gradients of functions of several variables, polar coordinates and the formal definition of a vector space. From linear algebra we will use eigenvalues and eigenvectors. We will once or twice need complex numbers and Cramer's rule for calculating inverse matrices. In Section 2.4 we calculate the surface area of a parametrized surface.

We also make use of a few ideas that are not always part of the standard beginning mathematics curriculum. The geometry of ellipses is crucial. This is part of the subject of analytic geometry, which is often included in calculus texts. See, for example, Shenk [Sh, Chapter 10]. Antisymmetry (also known as skew-symmetry) plays a large role. We define antisymmetric forms in Chapter 2 and antisymmetric matrices in Chapter 6. Matrix exponentiation is important in Chapter 6. To review or learn the theory of exponentiation of matrices, see [Ap97, Sections 9.5 to 9.10] or [St, Section 5.4]. In Chapter 6 we will also need to differentiate matrix-valued functions of one variable. This is done by differentiating each entry and collecting the results into a matrix in the obvious way. Some of the exercises ask the reader to solve differential equations, but these exercises can be skipped if necessary. Likewise, exercises relying explicitly on the implicit function theorem can be skipped by readers unfamiliar with that theorem.

Finally, there are a few essential topics that are not in the standard introductory texts, but are simple enough to introduce here.

Definition 1 *The* Cartesian product *of two vector spaces, V and W is the vector space*

$$V \times W := \{(v, w) : v \in V \text{ and } w \in W\},$$

with addition defined by $(v_1, w_1) + (v_2, w_2) := (v_1 + v_2, w_1 + w_2)$ and scalar multiplication defined by $r(v, w) := (rv, rw)$ for every $r \in \mathbb{R}$, every $v, v_1, v_2 \in V$, and every w, w_1 and $w_2 \in W$.

For example, the Cartesian product $\mathbb{R}^2 \times \mathbb{R}^3$ is the same as $\mathbb{R}^5$ (up to a few inessential parentheses).

Exercise 1 *Show that the Cartesian product $V \times W$ satisfies the axioms of a vector space. Show that the Cartesian product is associative (if we are willing to erase parentheses).*

In Chapter 7 we will use the *trace* of a matrix, that is, the sum of the entries on the main diagonal. More explicitly, the trace is defined, for any $n \times n$ matrix A, to be

$$\text{tr}A := \sum_{i=1}^{n} A_{ii}.$$

Exercise 2 *Show that for any $m \times n$ matrix A and any $n \times m$ matrix B we have* $\text{tr}(AB) = \text{tr}(BA)$. *Also show that if A and B are matrix-valued functions of t,* *then* $\frac{d}{dt}\text{tr}(AB) = \text{tr}(\frac{dA}{dt}B) + \text{tr}(A\frac{dB}{dt})$, *where* $\frac{d}{dt}$ *acts entry by entry on matrices.*

In Chapter 5 and in the optional Chapter 3 we will require the notions of injective and surjective functions.

Definition 2 *A function f is* injective *(also known as* one-to-one*) if for every x and y in the domain of f we have that*

$$f(x) = f(y) \text{ implies } x = y.$$

In other words, each output value of an injective function corresponds to a unique input value. For example, the function $f : \mathbb{R} \to \mathbb{R}, x \mapsto x^3$ is injective because each cube of a real number has a unique real cube root. On the other hand, $g : \mathbb{R} \to \mathbb{R}, x \mapsto x^2$ is not injective because, for example, $g(-2) = 4 = g(2)$ but $-2 \neq 2$. Notice though that injectivity of a function depends not only on the formula that defines the function but also on the domain: the function $h : \mathbb{R}^+ \to \mathbb{R}, x \mapsto x^2$ is injective, since the square of each strictly positive real number has a unique strictly positive square root.

Exercise 3 *Let A be an $n \times m$ matrix. Then A defines a function $f : \mathbb{R}^m \to \mathbb{R}^n, x \mapsto Ax$. Show that f is injective if and only if the rank of A is m.*

Surjectivity is closely related to injectivity.

Definition 3 *A function $f : D \to S$ is* surjective *(also known as* onto*) if for every $s \in S$ there is a $d \in D$ such that $f(d) = s$.*

For example, the function f defined above is surjective because every real number is the cube of some real number. However, g is not surjective because not every real number is the square of a real number. For example, there is no real number x such that $x^2 = -1$. The function h fails to be surjective for the same reason. One might reasonably object that one can always arrange surjectivity by defining the target space (labeled S in Definition 3) to be the range of the function. However, this is sometimes more trouble than it is worth. People who value precision over conciseness usually specify the target space explicitly by saying *surjective onto S*.

Exercise 4 *Let A be an $n \times m$ matrix. Then A defines a function $f : \mathbb{R}^m \to \mathbb{R}^n, x \mapsto Ax$. Show that f is surjective if and only if the rank of A is n.*

The reader can find more examples and exercises involving Cartesian products, injectivity and surjectivity in any standard analysis book, including Bartle [Ba, Chapter 1].

1
The Two-Body Problem

The goal of this chapter is to derive Kepler's laws of planetary motion from Newton's laws of motion and gravitation. We will analyze the motion of two massive particles (such as the sun and Mars), assuming that all forces other than mutual gravitational attraction are negligible. We want to find explicit formulas allowing us to predict motions. Will Mars crash into the sun? Will comet Hale-Bopp return? If so, when? Our formulas will be of use to engineers, who may want to maneuver satellites or send a space probe that is "parked," i.e., orbiting a planet, off to orbit a different planet. During this chapter we will pretend to be physicists. We will make savvy use of conserved quantities and particular functional forms to pick coordinates that simplify our calculations.

Historically, Kepler's laws came first. Using data on the trajectory of Mars obtained by Tycho Brahe (in whose observatory he was an assistant), Johannes Kepler discovered in the early 1600s that:

1. Planets move on ellipses with the sun at one focus.

2. The line from the sun to a planet sweeps out equal areas in equal times.

3. The period τ of a planet's orbit is related to the length a of its semimajor axis by
$$\tau^2 = Ka^3,$$
where K is a constant, now known to be $\frac{4\pi^2}{GM_s}$, with M_s equal to the mass of the sun and G equal to the universal gravitational constant.

The first and second laws were published in [Ke1] and the third in [Ke2]. Kepler's extraordinary genius and persistence in culling these geometric laws from tables of numerical data are beautifully described in a biography by Arthur Koestler [Koe]. The laws are in part an expression of Kepler's religious belief. In his own words (from *Tertius interveniens*, as quoted in [Koe]):

> Thus God himself
> was too kind to remain idle
> and began to play the game of signatures
> signing his likeness unto the world:
> therefore I chance to think
> that all nature and the graceful sky are
> symbolized in the art of Geometria.

But the laws are also good modern science: precise mathematical statements that predict behavior.

Exercise 5 *Check that Kepler's three laws imply that the ellipse on which the planet moves, as well as the position of the planet at any time, is determined by the initial position and velocity of the planet. There is a solution requiring only the geometry of the ellipse and a smidgen of differential calculus.*
Note: the reader may wish to postpone solving this exercise, which is significantly harder than most of the others in the book.

Sir Isaac Newton's *Principia Mathematica Philosophiae Naturalis* [N] appeared in the late 1600s and made crucial use of Kepler's laws. But the *Principia* was broader in scope, to put it mildly. Newton introduced fundamental laws of force and motion, a law of gravitation that explained both motion of objects near the earth's surface and the motion of planets and moons, as well as differential calculus. He analyzed a host of physical phenomena, including the ocean tides, air resistance and of course planetary motion. It would be hard to overstate the grandeur of Newton's contribution to modern science.

We will need Newton's second law of motion, commonly known as $F = ma$: force equals mass times acceleration. We will use Newton's original formulation: force equals the rate of change of (linear) momentum. *Linear momentum*, which Newton calls "the quantity of motion," is equal to mass times velocity, and acceleration is the time rate of change of velocity, so the two formulations are equivalent. We will also use Newton's universal law of gravitation and his notion of the center of mass.

1.1 First Simplification

Consider two isolated, massive particles. We might as well study the system in coordinates that move with the center of mass of the system (but don't rotate). This is the first simplification. See Figure 1.1. Sitting at the center of mass while the particles hurtle through space is like sitting in a train moving at constant speed along a straight track – you can juggle just as well in the train as you can in the station. Note, however, that juggling is harder if you are spinning. Any coordinate system whose origin moves at constant velocity and whose orientation is not changing is an *inertial system* because Newton's second law holds in such a coordinate system. We choose coordinates moving with center of mass, as in Figure 1.2, in order to simplify the problem. Let us denote the positions of the particles in center-of-mass coordinates by column vectors:

$$\mathbf{r}_1 := \begin{pmatrix} r_{1x} \\ r_{1y} \\ r_{1z} \end{pmatrix} \qquad \mathbf{r}_2 := \begin{pmatrix} r_{2x} \\ r_{2y} \\ r_{2z} \end{pmatrix}.$$

We will denote the momenta of the particles by row vectors:

$$\mathbf{p}_1 := \begin{pmatrix} p_{1x} & p_{1y} & p_{1z} \end{pmatrix} := \left(m_1 \frac{d}{dt} \mathbf{r}_1 \right)^T$$

$$\mathbf{p}_2 := \begin{pmatrix} p_{2x} & p_{2y} & p_{2z} \end{pmatrix} := \left(m_2 \frac{d}{dt} \mathbf{r}_2 \right)^T.$$

This choice of notation deserves some explanation. First of all we want our notation to be consistent throughout this book. So, although we are pretending in this chapter to be physicists, we still use the mathematicians' notation that

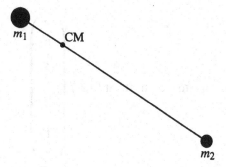

Figure 1.1. The distance from the center of mass to the particle with mass m_1 is $m_1/(m_1 + m_2)$ times the distance between the two particles. In this picture $m_1 > m_2$.

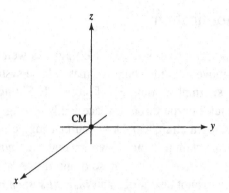

Figure 1.2. Center-of-mass coordinates

will be expedient later. Where physicists tend to stress the close relationship between quantities such as the position of a particle and its distance from the origin, mathematicians tend to stress the distinction between vector and scalar quantities as well as the functional dependence of one on the other. So if we (mathematicians) wish to discuss the distance of a particle from the origin, we will write $|\mathbf{r}|$, where $\mathbf{r}$ is the vector representing the position of the particle. The notation reminds us that distance depends on position, and indicates, more precisely, that distance is the absolute value of position. This differs from the standard physics notation, in which r might denote distance from the origin, while $\vec{r}$ or $\mathbf{r}$ denotes the position vector.

Our second remark on notation is that we will use row- and column-vectors to suggest a useful distinction that is obscured when one thinks only of column-vectors. We write momentum as a row-vector because *momentum is naturally dual to velocity*, in the sense that it often makes sense to perform the matrix multiplication

$$\begin{pmatrix} m & o & m & e & n & t & u & m \end{pmatrix} \begin{pmatrix} v \\ e \\ l \\ o \\ c \\ i \\ t \\ y \end{pmatrix},$$

which yields a scalar with units of energy. Although we could, as most physics texts do, write momentum as a column vector and use the dot product, we would rather discourage indiscriminate use of the dot product. This will help us avoid physically unnatural calculations (such as taking a dot product of two force vectors). Another example of a pair of naturally dual physical concepts is the relationship of force vectors to displacement vectors. To calculate work we multiply a force row-vector by a displacement column-vector to get a scalar (with the units of energy). Note that once we have chosen (arbitrarily) to use column-vectors for velocities, displacements should also be column-vectors, since velocity is the time derivative of displacement (and the derivative of a vector should be a vector of the same dimensions). Also, Newton's second law of motion tells us that force is the time derivative of momentum, so it is consistent to think of both as row-vectors. Following standard mathematical notation, we denote the vector space of column three-vectors by $\mathbb{R}^3$ and the space of row three-vectors by $(\mathbb{R}^3)^*$ (which can be pronounced "R-three-dual"). We discuss duality further in Section 2.2.

Using center-of-mass coordinates simplifies our calculations by giving two algebraic relations that allow us to reduce the number of variables in our differential equations. First of all, because the center of mass is given by

$$0 = CM = \frac{1}{m_1 + m_2}\left(m_1\mathbf{r}_1 + m_2\mathbf{r}_2\right),$$

we have $m_1\mathbf{r}_1 + m_2\mathbf{r}_2 = 0$. But we can differentiate both sides of this equation to learn that $\mathbf{p}_1 + \mathbf{p}_2 = \mathbf{0}$. In other words, using center-of-mass coordinates ensures not only that the center of mass is at the origin, but that the total linear momentum is zero.

Any results obtained in center-of-mass coordinates can easily be carried over to an arbitrary fixed inertial system. Let λ denote the (constant) total momentum of the two particles. Then the center of mass moves with constant velocity $\frac{\lambda}{m_1+m_2}$. To analyze the motion in this inertial system one merely combines this simple motion of the center of mass with the motion of the particles in center-of-mass coordinates. Specifically, to get position as a function of time in the arbitrarily chosen inertial system, one simply adds the position function in center-of-mass coordinates to the function giving the position of the center of mass (as a function of time) in the arbitrary inertial system.

Next we derive the differential equations of motion for $\mathbf{r}_1$, $\mathbf{r}_2$, $\mathbf{p}_1$ and $\mathbf{p}_2$ from Newton's second law of motion and his gravitational force law. Newton's second law of motion is commonly written $\mathbf{F} = m\mathbf{a}$ (force equals mass times acceleration), but we will think of it (as Newton did) as saying that force equals the rate of change of momentum:

$$\mathbf{F}_1 = \frac{d}{dt}\mathbf{p}_1 \text{ and } \mathbf{F}_2 = \frac{d}{dt}\mathbf{p}_2,$$

where $\mathbf{F}_1$ is the force experienced by the first particle and $\mathbf{F}_2$ is the force experienced by the second particle. Note that momentum is a physically natural quantity because it can be measured directly by experiment, namely, collision experiments. According to Newton's law of gravitation,

$$\mathbf{F}_1 = -G\frac{m_1 m_2}{|\mathbf{r}_1 - \mathbf{r}_2|^3}(\mathbf{r}_1 - \mathbf{r}_2)^T$$

$$\mathbf{F}_2 = -G\frac{m_1 m_2}{|\mathbf{r}_1 - \mathbf{r}_2|^3}(\mathbf{r}_2 - \mathbf{r}_1)^T .$$

(Experiments show that $G \approx 6.7 \times 10^{-11}\frac{Nm^2}{(kg)^2}$.) So Newton's laws, along with the definition of momentum, give a system of twelve ordinary differential equations. The first six scalar equations are

$$\left.\begin{array}{l} \frac{d}{dt}r_{1x} = \frac{p_{1x}}{m_1} \\[4pt] \frac{d}{dt}r_{1y} = \frac{p_{1y}}{m_1} \\[4pt] \frac{d}{dt}r_{1z} = \frac{p_{1z}}{m_1} \end{array}\right\} \text{ i.e., } \quad \frac{d\mathbf{r}_1}{dt} = \frac{\mathbf{p}_1^T}{m_1}$$

$$\left.\begin{array}{l} \frac{d}{dt}p_{1x} = -\frac{Gm_1 m_2(r_{1x}-r_{2x})}{|\mathbf{r}_1-\mathbf{r}_2|^3} \\[4pt] \frac{d}{dt}p_{1y} = -\frac{Gm_1 m_2(r_{1y}-r_{2y})}{|\mathbf{r}_1-\mathbf{r}_2|^3} \\[4pt] \frac{d}{dt}p_{1z} = -\frac{Gm_1 m_2(r_{1z}-r_{2z})}{|\mathbf{r}_1-\mathbf{r}_2|^3} \end{array}\right\} \text{ i.e., } \quad \frac{d\mathbf{p}_1^T}{dt} = -\frac{Gm_1 m_2(\mathbf{r}_1 - \mathbf{r}_2)}{|\mathbf{r}_1 - \mathbf{r}_2|^3}$$

(1.1)

and six more scalar equations come from

$$\frac{d\mathbf{r}_2}{dt} = \frac{\mathbf{p}_2^T}{m_2} \text{ and } \frac{d\mathbf{p}_2^T}{dt} = -\frac{Gm_1 m_2(\mathbf{r}_2 - \mathbf{r}_1)}{|\mathbf{r}_1 - \mathbf{r}_2|^3}. \tag{1.2}$$

These ordinary differential equations are *coupled* (because the derivatives of the $\mathbf{p}$'s depend on the $\mathbf{r}$'s and vice versa), *first-order* (because only first derivatives and no higher derivatives appear) and *nonlinear* (because, for example $-\frac{Gm_1 m_2(\mathbf{r}_2-\mathbf{r}_1)}{|\mathbf{r}_1-\mathbf{r}_2|^3}$ is not a linear function of the $\mathbf{r}$'s).

Newton's laws do not tell us how to *solve* these differential equations: they do not give explicit trajectories $\mathbf{r}_1(t)$ and $\mathbf{r}_2(t)$. They merely tell us what the differential equations are. In other words, they do not explicitly *predict* the motions, but they give us some rules that the motions must obey.

Let us exploit our first simplification, the passage to center-of-mass coordinates. Because of this simplification we have

$$\mathbf{p}_1 + \mathbf{p}_2 = 0 \text{ and } m_1\mathbf{r}_1 + m_2\mathbf{r}_2 = 0. \tag{1.3}$$

So we can specify the state of the system with the coordinates $\mathbf{r}$ and $\mathbf{p}$, where $\mathbf{r} = \mathbf{r}_1 - \mathbf{r}_2$ and $\mathbf{p} = \mathbf{p}_1$. Both $\mathbf{r}$ and $\mathbf{p}$ are three-vectors. Note that we can recover our original coordinates:

$$\mathbf{p}_1 = \mathbf{p}, \quad \mathbf{p}_2 = -\mathbf{p}, \quad \mathbf{r}_1 = \frac{m_2}{m_1 + m_2}\mathbf{r}, \quad \mathbf{r}_2 = \frac{-m_1}{m_1 + m_2}\mathbf{r}.$$

Geometrically, Equations 1.3 cut out a six-dimensional plane in twelve-dimensional *phase space* $\{(\mathbf{r}_1, \mathbf{r}_2, \mathbf{p}_1, \mathbf{p}_2)\} \cong \mathbb{R}^{12}$. Phase spaces are defined and discussed at length in Chapter 2. The new coordinates are coordinates on that plane.

What have we gained? Look what happens when we rewrite our system of twelve ordinary differential equations in the new coordinates:

$$\frac{d\mathbf{r}}{dt} = \frac{d\mathbf{r}_1}{dt} - \frac{d\mathbf{r}_2}{dt} = \frac{\mathbf{p}_1^T}{m_1} - \frac{\mathbf{p}_2^T}{m_2} = \left(\frac{1}{m_1} + \frac{1}{m_2}\right)\mathbf{p}^T$$

$$\frac{d\mathbf{p}^T}{dt} = \frac{d\mathbf{p}_1^T}{dt} = -\frac{Gm_1m_2(\mathbf{r}_1 - \mathbf{r}_2)}{|\mathbf{r}_1 - \mathbf{r}_2|^3} = -\frac{Gm_1m_2}{|\mathbf{r}|^3}\mathbf{r}.$$

We can interpret the system of six equations as equations of motion for a single particle, whose position in three-space is $\mathbf{r}$ and whose velocity is $\mathbf{p}$. This particle is sometimes called the *reduced particle* and its mass $\mu = \frac{m_1m_2}{m_1+m_2}$, which is smaller than either m_1 or m_2, is called the *reduced mass*. The reduced particle moves under the influence of the potential $-\frac{GM\mu}{|\mathbf{r}|}$, where $M = m_1 + m_2$ is the total mass of the system. This is called a *central potential* because it depends only on the length of $\mathbf{r}$. With this new notation the equations of motion are

$$\frac{d\mathbf{r}}{dt} = \frac{\mathbf{p}^T}{\mu} \text{ and } \frac{d\mathbf{p}^T}{dt} = -\frac{GM\mu}{|\mathbf{r}|^3}\mathbf{r}. \tag{1.4}$$

Notice that if we solve these six first-order ordinary differential equations we can recover the solution to the twelve original ordinary differential equations using our explicit formulas for $\mathbf{p}_1, \mathbf{p}_2, \mathbf{r}_1$ and $\mathbf{r}_2$ in terms of $\mathbf{p}$ and $\mathbf{r}$. This is the good news: we have *reduced* our twelve-dimensional system to a six-dimensional system. The bad news is that these equations are still coupled (because the derivative of $\mathbf{r}$ depends on $\mathbf{p}$ and vice versa) and nonlinear (because of the "$|\mathbf{r}|^3$" in the denominator). So it is still not clear how to solve them.

1.2 Second Simplification

So far we have reduced the two-body problem to the simpler problem of a single particle in a central potential. In this section we exploit the conservation of angular momentum to reduce the number of coordinates in our problem from six to four.

The equations of motion imply that the velocity vector $\frac{d\mathbf{r}}{dt}$ is parallel to the momentum vector $\mathbf{p}^T$ and the rate of change vector for the momentum $\frac{d\mathbf{p}^T}{dt}$ is parallel to the position vector $\mathbf{r}$. Hence if both $\mathbf{r}(t_0)$ and $\mathbf{p}(t_0)^T$ lie in a particular plane P, then for all time $\mathbf{r}$ and $\mathbf{p}^T$ lie in P. That is, if the position and velocity of our "reduced" particle lie in P at any one time, they stay in P for all time. But any two vectors determine a plane. So the motion for all time is contained in a fixed plane through the origin. (In the special case that $\mathbf{r}$ is parallel to $\mathbf{p}^T$ at any one time, the two vectors are parallel for all time and the motion is contained in a fixed line through the origin.) Another way to see that the motion is restricted to a plane is to note that

$$\frac{d}{dt}\left(\mathbf{r} \times \mathbf{p}^T\right) = \frac{d\mathbf{r}}{dt} \times \mathbf{p}^T + \mathbf{r} \times \frac{d\mathbf{p}^T}{dt} = \mathbf{0},$$

since $\frac{d\mathbf{r}}{dt}$ is parallel to $\mathbf{p}^T$ and $\frac{d\mathbf{p}^T}{dt}$ is parallel to $\mathbf{r}$. The constant vector $\mathbf{r} \times \mathbf{p}^T$ is called the *angular momentum vector*, and we define $\tilde{\mathbf{L}} = \mathbf{r} \times \mathbf{p}^T$. The reason for the tilde (on the $\tilde{\mathbf{L}}$) will become clear in Chapter 6. In the standard physics literature the angular momentum vector is denoted simply by $\mathbf{L}$. Recall from vector calculus that $\tilde{\mathbf{L}}$ must be perpendicular to the position vector $\mathbf{r}$. So the constancy of $\tilde{\mathbf{L}}$ (often called the *conservation of angular momentum* implies that the motion must lie in the plane through the origin. We will see in Section 1.3 that the constancy of $\tilde{\mathbf{L}}$ also implies Kepler's second law.

Exercise 6 *(Vector calculus review). Show that for any functions* $\mathbf{f}, \mathbf{g} : \mathbb{R} \to \mathbb{R}^3$,

  1.

$$\frac{d}{dt}(\mathbf{f}(t) \times \mathbf{g}(t)) = \mathbf{f}'(t) \times \mathbf{g}(t) + \mathbf{f}(t) \times \mathbf{g}'(t).$$

  2.

$$\frac{d}{dt}(\mathbf{f}(t) \cdot \mathbf{g}(t)) = \mathbf{f}'(t) \cdot \mathbf{g}(t) + \mathbf{f}(t) \cdot \mathbf{g}'(t).$$

3. *Show that for any vectors* **v** *and* **w** *in* $\mathbb{R}^3$, *the nonnegative scalar* $|\mathbf{v} \times \mathbf{w}|$ *is the area of the parallelogram spanned by* **v** *and* **w**.

We use the observation that the motion is restricted to a plane through the origin to choose convenient coordinates, namely, polar coordinates on that plane. (See Figure 1.3. Readers who would like to review polar coordinates can consult any standard multivariable textbook and most calculus textbooks; we recommend Spivak's beautiful text [Sp].) We can choose to orient our polar coordinates so that the vector **r** moves in the direction of increasing θ. (Because **r** is the position vector, we will use ρ for its length, i.e., the radial polar coordinate. Real physicists would happily denote both the vector and its length by the same letter, depending on context or typeface to show which is which.) Now we can describe the system using only four scalar coordinates, $\rho, \theta, \dot{\rho}$ and $\dot{\theta}$, defined by

$$\mathbf{r} = \rho \begin{pmatrix} \cos\theta \\ \sin\theta \\ 0 \end{pmatrix}$$

(1.5)

$$\frac{\mathbf{p}}{\mu} = \dot{\rho} \left(\cos\theta \quad \sin\theta \quad 0 \right) + \rho\dot{\theta} \left(-\sin\theta \quad \cos\theta \quad 0 \right),$$

with $\rho \geq 0$ and $\rho, \theta \in \mathbb{R}$. The right-hand rule for cross-products ensures counterclockwise motion (as seen from the $\tilde{\mathbf{L}}$ axis) and hence $\dot{\theta} \geq 0$.

It is important to think of $\dot{\rho}$ and $\dot{\theta}$ merely as coordinates, not as derivatives of ρ and θ. One can use Equations 1.5 to rewrite the equations of motion in the

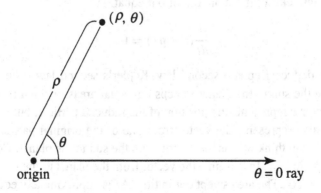

Figure 1.3. Polar coordinates

new coordinates, *reducing* our six-dimensional system to a four-dimensional system. The derivation is left as an exercise. The new equations are

$$2\dot{\rho}\dot{\theta} + \rho\frac{d\dot{\theta}}{dt} = 0$$

$$\frac{d\dot{\rho}}{dt} - \rho\dot{\theta}^2 + \frac{GM}{\rho^2} = 0$$

$$\frac{d\rho}{dt} = \dot{\rho} \tag{1.6}$$

$$\frac{d\theta}{dt} = \dot{\theta}$$

Exercise 7 *Derive these equations.*

The conservation of angular momentum and the observation that the motion lies in a plane has allowed us to simplify the two-body problem still further, to a system of first-order, nonlinear, coupled differential equations that is only four-dimensional. We will see in the next section that we can quickly derive Kepler's laws from this four-dimensional system.

1.3   Recovering Kepler's Laws

In this section we derive Kepler's laws from Equations 1.6, the analytic geometry of ellipses and the solution to one scalar, linear ordinary differential equation. We first derive Kepler's second law, then his first, and finally his third.

First we will eke a little more out of our equation

$$\frac{d}{dt}(\mathbf{r} \times \mathbf{p}^T) = \mathbf{0}$$

in order to deduce Kepler's second law. Kepler's second law states that the vector from the sun to the planet sweeps out equal areas in equal times. Note that the vector $\mathbf{r}$ represents the position of the reduced particle, but since $\mathbf{r} := \mathbf{r}_1 - \mathbf{r}_2$, it also represents the vector from one of the original particles to the other. So we can think of $\mathbf{r}$ as the vector from the sun to the planet. Consider a short time period Δt. In this time the vector from the planet to the sun changes from $\mathbf{r}$ to $\mathbf{r} + \Delta\mathbf{r}$. The area swept out in time Δt is approximately equal to the area of the triangle in Figure 1.4, namely, $\frac{1}{2}|\mathbf{r} \times \Delta\mathbf{r}|$. But $\Delta\mathbf{r}$ is approximately

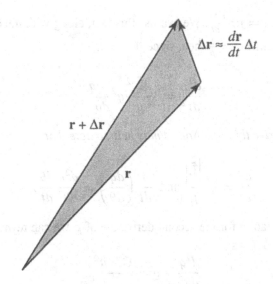

$$\Delta \mathbf{r} \approx \frac{d\mathbf{r}}{dt} \Delta t$$

$\mathbf{r} + \Delta \mathbf{r}$

$\mathbf{r}$

Figure 1.4.   Change in area in small time increment

equal to $\frac{d\mathbf{r}}{dt}\Delta t$ and, since $\mathbf{p}^T = \mu \frac{d\mathbf{r}}{dt}$ we conclude that the area swept out during time Δt is approximately

$$\frac{1}{2\mu} \left|\mathbf{r} \times \mathbf{p}^T\right| \Delta t = \frac{1}{2\mu} \left|\tilde{\mathbf{L}}\right| \Delta t.$$

The area swept out in any time period of length T is $\frac{1}{2\mu} \left|\tilde{\mathbf{L}}\right| T$. So the line from the sun to the planet sweeps out equal areas in equal times. We have recovered Kepler's second law.

Next we will use conservation of angular momentum and the Equations 1.6 to deduce Kepler's first law under the assumption that the angular momentum $\tilde{\mathbf{L}} \neq \mathbf{0}$. Combining three of the Equations 1.6 we obtain

$$\frac{d^2\rho}{dt^2} - \rho \left(\frac{d\theta}{dt}\right)^2 + \frac{GM}{\rho^2} = 0. \tag{1.7}$$

By the defining Equations (1.5) for $\rho, \theta, \dot{\rho}$ and $\dot{\theta}$, it is easy to check that

$$\tilde{\mathbf{L}} := \mathbf{r} \times \mathbf{p}^T = \begin{pmatrix} 0 \\ 0 \\ \mu\rho^2\dot{\theta} \end{pmatrix} = \begin{pmatrix} 0 \\ 0 \\ \mu\rho^2\frac{d\theta}{dt} \end{pmatrix},$$

where the last equality uses the last one of Equations 1.6. By our choice of orientation of polar coordinates on the plane of motion, the value of θ increases with time, so $|\tilde{\mathbf{L}}| = \mu \rho^2 \frac{d\theta}{dt}$. We can use this fact, along with the chain rule, to see that if we define $q = \frac{1}{\rho}$ then, since $|\tilde{\mathbf{L}}| \neq 0$,

$$\frac{d^2\rho}{dt^2} = -\frac{|\tilde{\mathbf{L}}|^2}{\mu^2} q^2 \frac{d^2 q}{d\theta^2}.$$

Exercise 8 *Derive this equation. It may help to note that*

$$\frac{d\theta}{dt} = q^2 \frac{|\tilde{\mathbf{L}}|}{\mu} \quad \text{and} \quad \frac{d}{dt}\left(\frac{dq}{d\theta}\right) = \frac{d^2 q}{d\theta^2} \frac{d\theta}{dt}.$$

Using the equation for the second derivative of ρ we can rewrite the differential equation (1.7) as

$$\frac{d^2 q}{d\theta^2} + q = \frac{GM\mu^2}{|\tilde{\mathbf{L}}|^2}.$$

This is a constant-coefficient, linear, inhomogeneous differential equation. One can use standard techniques from the undergraduate level theory of differential equations (explained in, e.g., [BD, Section 3.6] or [Ap69, Volume II, Section 7.8]) to see that every solution of this equation must be of the form

$$q = \frac{GM\mu^2}{|\tilde{\mathbf{L}}|^2} + A\cos(\theta + \theta_0),$$

where A and θ_0 are constants (of integration). Hence the polar coordinates of the vector from the sun to the planet satisfies the equation

$$\frac{1}{\rho} = \frac{GM\mu^2}{\left|\tilde{\mathbf{L}}\right|^2} + A\cos(\theta + \theta_0). \tag{1.8}$$

If one defines $x := \rho\cos(\theta + \theta_0)$ and $y := \rho\sin(\theta + \theta_0)$ and then performs algebraic manipulations, one can see that this gives a quadratic equation in x and y, so that the orbit of the planet must be a conic section with an axis of symmetry parallel to the ray $\theta = -\theta_0$ and one focus at the origin.

To see that the planets in our solar system travel on ellipses, note that their orbits are *bounded*, that is, no planet can move beyond the outer reaches of the solar system. So the orbit of the vector from the sun to the planet must be bounded. Because angular momentum is conserved, a planet cannot slow down too much, ruling out bounded motion along a parabola or hyperbola. So the vector from the sun to the planet traces out an ellipse with the origin at the one focus. Hence, relative to the sun, the planet itself traces out an ellipse with the sun at one focus. We have recovered Kepler's first law.

Finally, we will recover the third law from our explicit formula for the orbit (Equation 1.8), our derivation of Kepler's second law and the geometry of ellipses. Let τ denote the period of the planet's motion, i.e., the time it takes for the planet to go once around the sun. We know from our recovery of Kepler's second law that the area swept out by the planet in time τ is $\frac{1}{2}\tau\frac{|\tilde{L}|}{\mu}$. On the other hand, the area swept out by the planet in time τ is just the area of the ellipse, πab, where a and b are the lengths of the semimajor and semiminor axes, respectively. So

$$\frac{1}{2}\frac{\tau|\tilde{L}|}{\mu} = \pi ab.$$

Now, from the geometry of ellipses (Figure 1.5) we know that the height of the point on the ellipse above the focus is $\frac{b^2}{a}$, while from Equation 1.8 (noting that ρ attains its maximum and minimum values for $\theta + \theta_0 \in \pi\mathbb{Z}$, so $\theta + \theta_0 \in \frac{\pi}{2} + \mathbb{Z}\pi$ at the point above the focus) we find that

$$\frac{1}{(b^2/a)} = \frac{GM\mu^2}{|\tilde{L}|^2}.$$

Combining these last two equations we find

$$\tau^2 = \frac{4\pi^2 a^2 b^2 \mu^2}{|\tilde{L}|^2} = \frac{4\pi^2 a^3}{GM}.$$

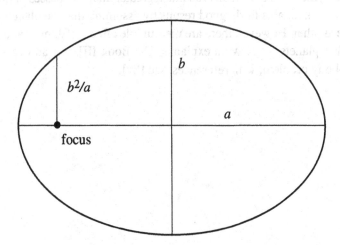

Figure 1.5. Geometry of an ellipse

Recall that $M := m_1 + m_2$, the sum of the masses. Because the sun is so much more massive than any planet we find for planets that

$$\tau^2 = \frac{4\pi^2 a^3}{GM} \approx \frac{4\pi^2 a^3}{GM_s},$$

where M_s is the mass of the sun. So we have recovered Kepler's third law.

Exercise 9 *Will Mars ever crash into the sun?*

Exercise 10 *Will comet Hale–Bopp return? When? According to the website of the Harvard-Smithsonian Center for Astrophysics [HS], when the comet was closest to the sun (0.91 AU from the sun) in early April 1997, it was traveling at about 44 km/sec (27 miles/sec). One astronomical unit (AU) is the average distance between the Sun and Earth, or about 93 million miles.*

Exercise 11 *Consider a satellite in a circular orbit above Earth's equator. Find the velocity of the satellite in terms of the distance a of the satellite from the center of the Earth. Then find the change in velocity needed to maneuver the satellite into a circular orbit that passes over the north and south poles.* (For the solution and other related problems, see Szebehely [Sz, Chapter 8].)

To be honest, we must remark that we have recovered Kepler's laws only for a solar system with one planet. In fact, an exact analysis in *closed form* (i.e., a formula for the positions of the bodies in terms only of t and familiar functions) of the gravitational interaction of more than two bodies is impossible. In our solar system, the mass of the sun is so much greater than the masses of the other planets that one obtains fairly good results by assuming that the planets do not affect one another. However, there are measurable effects of Jupiter and Saturn on the other planets, as Newton explained [N, Book III]. For an overview of the three-body problem, with references, see [Sz].

2

Phase Spaces of Mechanical Systems are Symplectic Manifolds

The concept, if not the name, of a symplectic manifold is familiar to any student of mechanics. Each symplectic manifold we study in this chapter is a *phase space* of a mechanical system. In other words, each is a set of distinguishable *states* of a particular mechanical system. For instance, to specify the state of a particle in space, it is enough to specify its position and momentum. Because Newton's second law $\mathbf{F} = m\mathbf{a}$ is second-order (acceleration $\mathbf{a}$ is the second derivative of position with respect to time, and the force $\mathbf{F}$ can depend on position and velocity but not on acceleration), knowing the position and velocity at any one time allows one to predict the particle's motion at all future times. To put it another way, if we think of Newton's law as $\mathbf{F} = \frac{d}{dt}\mathbf{p}$ and remember that the force never depends on derivatives of the momentum $\mathbf{p}$, we see that knowing the position and momentum allows one to predict the particle's motion at all future times. So we can think of the phase space as made up of position-momentum pairs. (Some authors use the phrase "state space" synonymously with "phase space.") Note that the *configuration space* is something different: it is the set of possible positions.

But phase space is not merely a set of points; it is a set of points with certain very nice coordinate systems called *canonical coordinates*. Canonical coordinates are coordinates in which the equations of motion for any classical dynamics problem have a certain simple form (given in general in Exercise 64 and for the two-dimensional version in Equations 4.1 in Section 4.1). For a discussion of canonical coordinates from a physicist's point of view, see Goldstein [Go, Chapter 8] or Török [To, p. 263]. For a mathematical treatment, see Abraham and Marsden [AM, Section 3.2]. One can think of the canonical coordinates as encoding the *kinematics* of the physical problem, i.e., the features of the motion that do not depend on the particulars of the force. We will encode all the information about canonical coordinates in an object called a *differential two-form*, which we will define in Definition 7 of Section 2.3.

Clearly kinematics alone is unsatisfying — we do want to consider particular problems with particular forces, and we will do so in Chapter 4, when we introduce the force by introducing the Hamiltonian energy function. Once we have specified forces, we can analyze the *dynamics*, i.e., the particular behavior. In the current chapter, however, we concentrate on kinematics.

In summary, one can think of a symplectic manifold as a set of points with all the necessary structure for kinematics. We denote symplectic manifolds by pairs (M, ω), where M is a set of points and ω is a differential two-form encoding the kinematics. Not every differential two-form is symplectic: we give the formal, correct mathematical definition in Definition 15 in Section 3.3.

The goal of this chapter is to present a few examples of symplectic manifolds, as well as to introduce some of the formalism of differential forms that we will need in our mathematical treatment of the two-body problem. In particular, we will learn about vector fields and differential forms, and in Section 2.6 we will see how to rewrite differential forms after changing coordinate systems – in other words, we will see how to pull a differential form back under a map from the manifold to itself.

2.1 The Plane and the Area Form

Our first example of a symplectic manifold is the two-dimensional plane with the area form. This is the phase space of a particle constrained to move on a line. We write

$$\mathbb{R}^2 = \left\{ \begin{pmatrix} r \\ p \end{pmatrix} : r \in \mathbb{R}, p \in \mathbb{R} \right\}$$

and introduce the bilinear form

$$A : \qquad \mathbb{R}^2 \times \mathbb{R}^2 \to \mathbb{R}$$
$$\left(\begin{pmatrix} r_1 \\ p_1 \end{pmatrix}, \begin{pmatrix} r_2 \\ p_2 \end{pmatrix} \right) \mapsto r_1 p_2 - r_2 p_1.$$

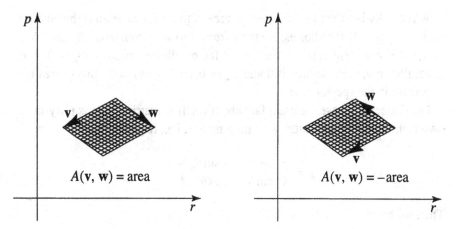

Figure 2.1. The area form

This is the phase space of a particle on a line if we interpret r as the position of the particle and p as its momentum. These two coordinates have a simple kinematic relationship since p is a constant multiple of the time derivative of r for any physically feasible motion. In fact, the coordinates r and p are canonical coordinates.

The function A is called the *area form* because $|A(\mathbf{v}, \mathbf{w})|$ gives the area of the parallelogram spanned by the vectors $\mathbf{v}$ and $\mathbf{w}$. We can use the tools of linear algebra to study A if we notice that

$$r_1 p_2 - r_2 p_1 = \det \begin{pmatrix} r_1 & r_2 \\ p_1 & p_2 \end{pmatrix} = \begin{pmatrix} r_1 & p_1 \end{pmatrix} \begin{pmatrix} 0 & 1 \\ -1 & 0 \end{pmatrix} \begin{pmatrix} r_2 \\ p_2 \end{pmatrix}.$$

Exercise 12 *Fix a nonzero vector* $\mathbf{v} \in \mathbb{R}^2$*. For what vectors* $\mathbf{w}$ *do we have* $A(\mathbf{v}, \mathbf{w}) = 1$*? Draw* $\mathbf{v}$ *and the set of possible* $\mathbf{w}$*'s.*

The sign of $A(\mathbf{v}, \mathbf{w})$ depends on the relative orientation of $\mathbf{v}$ and $\mathbf{w}$. See Fig. 2.1.

Proposition 1 *The number* $A(\mathbf{v}, \mathbf{w})$ *gives the signed area of the parallelogram spanned by* $\mathbf{v}$ *and* $\mathbf{w}$*, where the sign is positive if the shortest rotation from* $\mathbf{v}$ *to* $\mathbf{w}$ *is counterclockwise, negative if the shortest rotation is clockwise, and zero if* $\mathbf{v}$ *and* $\mathbf{w}$ *are parallel.*

Proof. First we attack a special case. Then we will use that case to prove the proposition in general.

As a special case, we consider vectors $\mathbf{v}$ pointing in the direction of the positive r-axis. That is, assume $\mathbf{v} = (c, 0)^T$ for some $c \in \mathbb{R}$, with $c \geq 0$. Then we have $A(\mathbf{v}, \mathbf{w}) = c w_2$ for any $\mathbf{w} = (w_1, w_2)^T$. If the shortest rotation from $\mathbf{v}$

to $\mathbf{w}$ is clockwise, then $w_2 < 0$ and the area of the parallelogram is the positive number $-cw_2$. If the shortest rotation from $\mathbf{v}$ to $\mathbf{w}$ is counterclockwise, then $A(\mathbf{v}, \mathbf{w}) = cw_2$, which is just the area of the parallelogram since $w_2 > 0$. If $\mathbf{w}$ is parallel to v, then the area is 0 and $w_2 = 0$, so $A(\mathbf{v}, \mathbf{w}) = 0$. This proves the statement in the special case.

For the general case we use a fact about rotation matrices that is nifty in its own right. Let g be an arbitrary rotation matrix, i.e., for some θ,

$$g = \begin{pmatrix} \cos\theta & -\sin\theta \\ \sin\theta & \cos\theta \end{pmatrix}.$$

Then we have

$$g^T \begin{pmatrix} 0 & 1 \\ -1 & 0 \end{pmatrix} g = \begin{pmatrix} 0 & 1 \\ -1 & 0 \end{pmatrix}.$$

One can prove this fact algebraically by matrix multiplication, or prove it geometrically by noting that g rotates the plane counterclockwise through the angle θ, g^T rotates it clockwise through the angle θ and the matrix on the right-hand side rotates it clockwise through the angle $\pi/2$.

We now use what we have learned about rotation matrices to prove the general case of our statement. Let $\mathbf{v}$ and $\mathbf{w}$ be arbitrary vectors in $\mathbb{R}^2$. Let g be a rotation matrix such that $g\mathbf{v}$ points in the direction of the positive r-axis. Then

$$A(\mathbf{v}, \mathbf{w}) = \mathbf{v}^T \begin{pmatrix} 0 & 1 \\ -1 & 0 \end{pmatrix} \mathbf{w} = \mathbf{v}^T g^T \begin{pmatrix} 0 & 1 \\ -1 & 0 \end{pmatrix} g\mathbf{w} = A(g\mathbf{v}, g\mathbf{w}),$$

which (by the special case) equals the signed area of the parallelogram spanned by $g\mathbf{v}$ and $g\mathbf{w}$, which equals the signed area of the parallelogram spanned by $\mathbf{v}$ and $\mathbf{w}$, since rotating two vectors simultaneously does not change the signed area of the parallelogram they span. So we have proved that for all vectors $\mathbf{v}, \mathbf{w} \in \mathbb{R}^2$, the number $A(\mathbf{v}, \mathbf{w})$ is the signed area of the parallelogram spanned by $\mathbf{v}$ and $\mathbf{w}$. This ends the proof of Proposition 1. $\Diamond$

The kind of argument made in this proof is common in both mathematics and physics, but the two disciplines use different words to describe it. A mathematician might say that we have used the *invariance* of the area form under rotation (i.e., the fact that $A(\mathbf{v}, \mathbf{w}) = A(g\mathbf{v}, g\mathbf{w})$) while a physicist could express the same idea by remarking that we should choose a convenient coordinate system for our calculation, namely, the coordinate system in which the vector $\mathbf{v}$ is parallel to the first coordinate axis. The physicist is implicitly using the intuition that calculations about areas can be carried out equally well in any orthogonal coordinate system.

By the way, orthogonal coordinate systems are not the only ones in which areas can be measured consistently. In other words, rotations are not the only linear transformations preserving area.

Exercise 13 *Find a necessary and sufficient condition on the entries of a* 2×2 *matrix for that matrix to preserve the area form.*

Exercise 14 *The product* $\mathbf{u} \cdot (\mathbf{v} \times \mathbf{w})$ *is called the* triple scalar product *of* $\mathbf{u}$, $\mathbf{v}$ *and* $\mathbf{w}$.

1. *Show that* $\mathbf{u} \cdot (\mathbf{v} \times \mathbf{w})$ *is the determinant of the matrix with columns* $\mathbf{u}$, $\mathbf{v}$ *and* $\mathbf{w}$.

2. *Interpret* $\mathbf{u} \cdot (\mathbf{v} \times \mathbf{w})$ *geometrically in terms of the parallelepiped spanned by* $\mathbf{u}$, $\mathbf{v}$ *and* $\mathbf{w}$.

3. *Show that* $\mathbf{u} \cdot (\mathbf{v} \times \mathbf{w}) = \mathbf{v} \cdot (\mathbf{w} \times \mathbf{u}) = \mathbf{w} \cdot (\mathbf{u} \times \mathbf{v})$ *for all vectors* $\mathbf{u}$, $\mathbf{v}$, $\mathbf{w} \in \mathbb{R}^3$, *by direct computation or by using 1 or 2.*

4. *Show that a* 3×3 *matrix M has determinant 1 if and only if the function* $\mathbb{R}^3 \to \mathbb{R}^3$, $v \mapsto Mv$ *takes the unit cube in the domain to a parallelepiped of signed volume 1.*

Exercise 15 *Show that, for any natural number n, an* $n \times n$ *matrix M has determinant 1 if and only if the function* $\mathbb{R}^n \to \mathbb{R}^n$, $v \mapsto Mv$ *takes the unit n-cube in the domain to a parallelepiped of signed n-dimensional volume 1.* **Hint:** If A is invertible it can be decomposed into a product of elementary matrices, i.e., matrices corresponding to elementary row operations. See any linear algebra textbook, such as [L, Section 2.2], or [HH, Section 2.3].

The two-dimensional plane with the area form is the phase space for any mechanical system consisting of one particle constrained to move in a one-dimensional configuration space. We might think, for example, of a bead on a straight wire, or a mass on a spring that doesn't bend or rotate. In this interpretation, r represents the position of the particle and p represents the momentum of the particle. If we imagine any trajectory in phase space that corresponds to an actual motion of a particle of constant mass, the position as a function of time determines the momentum. So these two coordinates, which are independent coordinates on phase space, have a special kinetic relationship with one another. That relationship is expressed mathematically by the area form.

2.2 Vectors, Covectors and Antisymmetric Bilinear Forms

Next we develop some of the theory of vectors and covectors. As an application of our new technology we will show that we can rewrite the area form A as $-dp \wedge dr$. First we introduce new notation for vectors parallel to $\mathbb{R}^2$. We write

$$\frac{\partial}{\partial r} := \begin{pmatrix} 1 \\ 0 \end{pmatrix} \text{ and } \frac{\partial}{\partial p} := \begin{pmatrix} 0 \\ 1 \end{pmatrix}.$$

Thus we can rewrite an arbitrary vector $(v_1, v_2)^T$ as $v_1 \frac{\partial}{\partial r} + v_2 \frac{\partial}{\partial p}$. Notice that we are taking care to write elements of $\mathbb{R}^2$ as column vectors.

The use of partial derivative notation here is not frivolous: vectors correspond naturally to *partial differential operators*. In other words, any vector on the plane suggests a way of differentiating functions of two variables; for instance, a unit vector suggests taking a directional derivative. It is often enlightening and never misleading to interpret a vector tangent to a space as a partial differential operator on functions on that space. For example, $\frac{\partial}{\partial p}$ operates on any function f of r and p, producing the partial derivative $\frac{\partial f}{\partial p}$. Because it is natural to add partial derivatives and multiply them by scalars (e.g., $2\frac{\partial f}{\partial r} + 3\frac{\partial f}{\partial p}$ makes perfect sense), it is natural to add partial differential operators and multiply them by scalars (the expression $2\frac{\partial}{\partial r} + 3\frac{\partial}{\partial p}$ makes perfect sense too). So partial differential operators naturally have the structure of a vector space. In Section 2.3 we will study partial differential operators whose coefficients may vary (a.k.a. *vector fields*), but for now we consider only constant coefficients (such as 2 and 3 above).

Exercise 16 *Use the partial differential operator interpretation to show that $\frac{\partial}{\partial r}$ and $\frac{\partial}{\partial p}$ are linearly independent.*
Hint: A partial differential operator is 0 if it evaluates to 0 on any differentiable function.

Next we introduce complementary notation for the *dual space* of $\mathbb{R}^2$. The dual space, denoted $(\mathbb{R}^2)^*$, is the set of linear, real-valued functions on $\mathbb{R}^2$. By a theorem of linear algebra (see, e.g., [L, Thm. 10, p. 77]), each linear, real-valued function on $\mathbb{R}^2$ can be written as a 1×2 vector; in other words, we can think of the elements of $(\mathbb{R}^2)^*$ as as row vectors. We give special names to the standard basis elements:

$$dr := \begin{pmatrix} 1 & 0 \end{pmatrix} \text{ and } dp := \begin{pmatrix} 0 & 1 \end{pmatrix}.$$

We call this the *dual basis to* $\{\frac{\partial}{\partial r}, \frac{\partial}{\partial p}\}$ because of the relations

$$dr(\tfrac{\partial}{\partial r}) = \begin{pmatrix} 1 & 0 \end{pmatrix} \begin{pmatrix} 1 \\ 0 \end{pmatrix} = 1 \quad dr(\tfrac{\partial}{\partial p}) = \begin{pmatrix} 1 & 0 \end{pmatrix} \begin{pmatrix} 0 \\ 1 \end{pmatrix} = 0$$

$$dp(\tfrac{\partial}{\partial r}) = \begin{pmatrix} 0 & 1 \end{pmatrix} \begin{pmatrix} 1 \\ 0 \end{pmatrix} = 0 \quad dp(\tfrac{\partial}{\partial p}) = \begin{pmatrix} 0 & 1 \end{pmatrix} \begin{pmatrix} 0 \\ 1 \end{pmatrix} = 1.$$

Any (constant-coefficient) linear combination of dr and dp is called a *covector*. (When we want to let the coefficients depend on the variables r and p we will have to talk about differential one-forms; see Section 2.3.) The choice of notation is properly suggestive: covectors are related to variables of integration and the differentials that make a cameo appearance in many calculus books. For now, the reader should not let the notation obscure the fact that a covector is simply a function — in fact, dr is the function $\begin{pmatrix} r \\ p \end{pmatrix} \mapsto r$ and dp is the function $\begin{pmatrix} r \\ p \end{pmatrix} \mapsto p$. A useful way to visualize a covector on $\mathbb{R}^2$ is to visualize its graph (a nonvertical plane through the origin in $\mathbb{R}^3$, the Cartesian product of the domain and the range of the covector) or its level curves (evenly spaced parallel lines in the domain, $\mathbb{R}^2$, as in Figure 2.2). Then one can visualize geometrically the duality with vectors in the domain: superimpose the vector on the level sets and ask how much the function changes from the base point of the vector to the tip.

What shape does duality take if we insist on thinking of vectors as first-order partial differential operators? Because the covectors are linear functions from $\mathbb{R}^2$ to $\mathbb{R}$, their first-order partial derivatives are (constant) scalars! So

$$(\alpha_1 dr + \alpha_2 dp)(v_1 \frac{\partial}{\partial r} + v_2 \frac{\partial}{\partial p}) = (v_1 \frac{\partial}{\partial r} + v_2 \frac{\partial}{\partial p})(\alpha_1 r + \alpha_2 p)$$

$$= v_1 \alpha_1 + v_2 \alpha_2 = \begin{pmatrix} \alpha_1 & \alpha_2 \end{pmatrix} \begin{pmatrix} v_1 \\ v_2 \end{pmatrix}.$$

For readers familiar with expressions such as $2dr + 9dp$ from the context of integration in multivariable calculus (see, e.g., [MTW, p. 387]), it is useful to understand that these kinds of integrands on $\mathbb{R}^2$ are naturally dual to vectors tangent to $\mathbb{R}^2$: at any particular point $(r_0, p_0) \in \mathbb{R}^2$ there is a natural scalar product of a tangent vector $\mathbf{v}$ at (r_0, p_0) with a covector α, namely, the line integral of the covector α over the straight line segment starting at (r_0, p_0) and having length and direction given by the vector $\mathbf{v}$. More precisely, a vector $\mathbf{v} = (v_1, v_2)^T$ and a covector $\alpha = \alpha_1 dr + \alpha_2 dp$ at a point $\mathbf{q} = (r_0, p_0)^T$

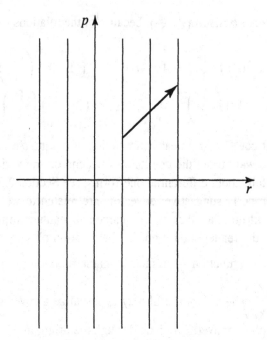

Figure 2.2. Level sets of the covector dr

naturally yield the scalar given by

$$\frac{d}{dt}\Big|_{t=0} \int_{\mathbf{q}}^{\mathbf{q}+t\mathbf{v}} \alpha = \frac{d}{dt}\Big|_{t=0} \int_{(r_0,p_0)}^{(r_0+tv_1,\,p_0+tv_2)} \alpha_1 dr + \alpha_2 dp.$$

Exercise 17 *Show that*

$$\frac{d}{dt}\Big|_{t=0} \int_{\mathbf{q}}^{\mathbf{q}+t\mathbf{v}} \alpha = \alpha_1 v_1 + \alpha_2 v_2.$$

The relationship between displacement, force and work is a physical expression of duality. One can think of a constant force as a covector, since it is a linear function from displacements (which are vectors) to work (which is a scalar). There is a similar relationship between velocity, momentum and kinetic energy. It is interesting to note that in both cases the scalar quantity has the units of energy.

The final animal in our constant-coefficient menagerie is the *antisymmetric bilinear form*. A *bilinear form* on $\mathbb{R}^2$ is a function $F : \mathbb{R}^2 \times \mathbb{R}^2 \to \mathbb{R}$ that is linear in each variable separately. That is, if we fix a point $\mathbf{c} \in \mathbb{R}^2$, then the functions $F(\mathbf{v}, \mathbf{c})$ and $F(\mathbf{c}, \mathbf{v})$ must both be linear functions of $\mathbf{v} \in \mathbb{R}^2$. To write it a different way, for each $\mathbf{c} \in \mathbb{R}^2$, both $F(\cdot, \mathbf{c})$ and $F(\mathbf{c}, \cdot)$ must be linear, real-valued functions on $\mathbb{R}^2$. In other words, both $F(\cdot, \mathbf{c})$ and $F(\mathbf{c}, \cdot)$ must be elements of the dual space $(\mathbb{R}^2)^*$ in order for F to be a bilinear form. A bilinear form F is *antisymmetric* if it satisfies $F(\mathbf{v}, \mathbf{w}) = -F(\mathbf{w}, \mathbf{v})$ for all $\mathbf{v}, \mathbf{w} \in \mathbb{R}^2$. An antisymmetric bilinear form F is *nondegenerate* if $F(\mathbf{c}, \cdot)$ is nonzero for every $\mathbf{c} \in \mathbb{R}^2$. In other words, nondegeneracy means that for every $\mathbf{c} \in \mathbb{R}^2$ there is a $\mathbf{v} \in \mathbb{R}^2$ such that $F(\mathbf{c}, \mathbf{v}) \neq 0$.

One way to build antisymmetric bilinear forms is to take *wedge products* of covectors.

Definition 4 *If α and β are covectors in $(\mathbb{R}^2)^*$, then the* wedge product of α *and β, denoted $\alpha \wedge \beta$, is defined by*

$$\alpha \wedge \beta : \qquad \mathbb{R}^2 \times \mathbb{R}^2 \to \mathbb{R}$$
$$(\mathbf{v}, \mathbf{w}) \mapsto (\alpha\mathbf{v})(\beta\mathbf{w}) - (\alpha\mathbf{w})(\beta\mathbf{v}).$$

It is easy to see that the wedge product is bilinear, since $(\alpha \wedge \beta)(\mathbf{w}, \mathbf{v})$ is linear in $\mathbf{v}$ when we hold $\mathbf{w}$ fixed, and vice versa. Also $\alpha \wedge \beta$ is antisymmetric, since switching $\mathbf{v}$ and $\mathbf{w}$ produces a minus sign:

$$(\alpha \wedge \beta)(\mathbf{w}, \mathbf{v}) = (\alpha\mathbf{w})(\beta\mathbf{v}) - (\alpha\mathbf{v})(\beta\mathbf{w}) \qquad (2.1)$$
$$= -(\alpha\mathbf{v})(\beta\mathbf{w}) + (\alpha\mathbf{w})(\beta\mathbf{v}) = -(\alpha \wedge \beta)(\mathbf{v}, \mathbf{w}). \qquad (2.2)$$

So the wedge product of any two covectors is an antisymmetric bilinear form.

Exercise 18 *Show that there is a one-to-one correspondence between antisymmetric bilinear forms on $\mathbb{R}^2$ and constant, antisymmetric 2×2 matrices.*

Note that $dp \wedge dr$ is the same as our area form A (up to a sign) since

$$
\begin{aligned}
(dp \wedge dr)(\mathbf{v}, \mathbf{w}) &= (dp(\mathbf{v}))(dr(\mathbf{w})) - (dp(\mathbf{w}))(dr(\mathbf{v})) \\
&= \left(dp(v_1\frac{\partial}{\partial r} + v_2\frac{\partial}{\partial p})\right)\left(dr(w_1\frac{\partial}{\partial r} + w_2\frac{\partial}{\partial p})\right) \\
&\quad - \left(dp(w_1\frac{\partial}{\partial r} + w_2\frac{\partial}{\partial p})\right)\left(dr(v_1\frac{\partial}{\partial r} + v_2\frac{\partial}{\partial p})\right) \\
&= v_2 w_1 - w_2 v_1 = -A(\mathbf{v}, \mathbf{w}).
\end{aligned}
$$

Exercise 19 *Show that the wedge product is itself antisymmetric, i.e., for any covectors α and β we have $\alpha \wedge \beta = -\beta \wedge \alpha$. In particular, we can write $A = dr \wedge dp$.*

Exercise 20 *Show that every antisymmetric bilinear form on $\mathbb{R}^2$ can be written as a wedge product of two covectors.*

It is not hard to see that the ideas introduced in this section can be generalized from $\mathbb{R}^2$ to any $\mathbb{R}^n$. We can think of $\mathbb{R}^n$ as column n-vectors or as linear, first-order partial differential operators. The dual space $(\mathbb{R}^n)^*$ consists of covectors, which can be thought of as row-vectors or as constant-coefficient differentials in n variables. Antisymmetric bilinear forms and wedge products are defined exactly as above, only now they are functions from $\mathbb{R}^n \times \mathbb{R}^n$ to $\mathbb{R}$. However, while the technology generalizes to higher dimensions, some of the results of this section are special to $\mathbb{R}^2$. In higher dimensions the volume form is not a bilinear form (it is a function of more than two variables), and not every antisymmetric bilinear form can be written as a wedge product.

Exercise 21 *Show that every antisymmetric bilinear form on $\mathbb{R}^3$ is a wedge product of two covectors.*

Exercise 22 *Find an antisymmetric bilinear form on $\mathbb{R}^4$ that cannot be written as a wedge product of two covectors. Equivalently, find a 4×4 antisymmetric matrix of rank 4. Show that every antisymmetric bilinear form on $\mathbb{R}^4$ can be written as a finite linear combination of wedge products of covectors.*

Also, in the n-dimensional picture corresponding to Figure 2.2, each level set is an *(n − 1)-dimensional hyperplane*, i.e., a set of all solutions to a single linear scalar equation in n variables. To tackle the two-body problem we will need to consider $\mathbb{R}^{12}$, the kinetic phase space for a two-particle system in three-dimensional space. In the next exercise, we ask the reader to consider the phase space of the two-body problem in center-of-mass coordinates:

Exercise 23 *Consider $\mathbb{R}^6$ with coordinates $r_x, r_y, r_z, p_x, p_y, p_z$. Find the constant 6×6 matrix corresponding to the antisymmetric bilinear form $dr_1 \wedge dp_1 + dr_2 \wedge dp_2 + dr_3 \wedge dp_3$.*

Exercise 24 *Show that there is a one-to-one correspondence between antisymmetric bilinear forms on $\mathbb{R}^n$ and constant, antisymmetric $n \times n$ matrices.*

There is another direction of generalization. We have seen covectors (linear forms) and bilinear forms; what about trilinear forms, quadrilinear forms and so on? These exist and are useful in many contexts. In this book we will not need them, except for a brief application of *alternating trilinear forms* in Section 3.3. A *trilinear form* on $\mathbb{R}^n$ is a function $F : \mathbb{R}^n \times \mathbb{R}^n \times \mathbb{R}^n \to \mathbb{R}$ that is linear in each variable separately. A trilinear form F is *alternating* if it satisfies $F(\mathbf{u}, \mathbf{v}, \mathbf{w}) = -F(\mathbf{v}, \mathbf{u}, \mathbf{w}) = F(\mathbf{v}, \mathbf{w}, \mathbf{u})$ for all vectors $u, v, w \in \mathbb{R}^n$. One way to build alternating trilinear forms is to take wedge products of covectors and antisymmetric bilinear forms.

Definition 5 *If α is a covector and β is an antisymmetric bilinear form, then the wedge product of α and β, denoted $\alpha \wedge \beta$, is defined by*

$$\alpha \wedge \beta : \qquad\qquad\qquad \mathbb{R}^n \times \mathbb{R}^n \times \mathbb{R}^n \to \mathbb{R}$$

$$(\mathbf{u}, \mathbf{v}, \mathbf{w}) \mapsto (\alpha(\mathbf{u}))(\beta(\mathbf{v}, \mathbf{w})) + (\alpha(\mathbf{v}))(\beta(\mathbf{w}, \mathbf{u})) + (\alpha(\mathbf{w}))(\beta(\mathbf{u}, \mathbf{v})).$$

Exercise 25 *Show that the trilinear form $\alpha \wedge \beta$ is alternating.*

All these ideas can be generalized to curved spaces. The first step toward this generalization is to consider varying coefficients. That is, instead of considering vectors, covectors and matrices with constant entries, we must let the entries vary with the coordinates. We take this first step in Section 2.3 and consider a curved space (the cylinder) in Section 2.4. In Section 3.3 we extend these ideas to a large class of curved spaces, manifolds.

2.3   Vector Fields and Differential Forms

Physicists and mathematicians tend to have different attitudes toward generalizing concepts. Physicists are usually willing to plow ahead, subjecting new notation to old rules of computation and intuiting new rules as necessary. Mathematicians usually progress more carefully, specifying the domain of every function in sight and keeping a strict list of which calculation rules are valid. To introduce vector fields and differential forms physics style, we simply say that a vector field is a vector with coefficients that may vary; a differential one-form is a covector with coefficients that may vary and a differential two-form is an antisymmetric bilinear form with coefficients that may vary. In this section we will first give some examples and then give a more mathematical approach to these objects.

Consider, for example, the vector field $(-p, r)^T$ on $\mathbb{R}^2$, shown in Figure 2.3. Note that the coefficients $-p$ and r do indeed vary: they are not constant as functions of r and p. We will see a slight variant of this vector field in Section 4.2 when we define and study the *Hamiltonian vector field* for a particle moving on a line under the influence of a spring.

Another example of a vector field is $\frac{\partial}{\partial \theta}$, where θ is the angle variable in polar coordinates (see Figure 1.3 in Section 1.2). Since we are using r as a Cartesian coordinate, we will use the Greek letter rho (ρ) to denote the distance from the origin. (We advise the reader to distinguish ρ from p in the mind and in calligraphy.) A point with polar coordinates $(\rho, \theta)^T$ corresponds to a point with Cartesian coordinates $(r = \rho \cos\theta, p = \rho \sin\theta)^T$. We define $\frac{\partial}{\partial \theta}$ at a point to be the velocity vector $\frac{\partial}{\partial \theta}(\rho \cos\theta, \rho \sin\theta)^T$ at the point in question.

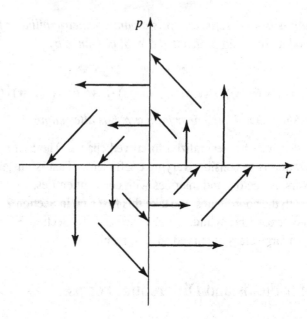

Figure 2.3. The vector field $(-p, r)^T$, also known as $-p\frac{\partial}{\partial r} + r\frac{\partial}{\partial p}$

Exercise 26 *Carry out the suggested differentiation to show that*

$$\frac{\partial}{\partial \theta} = \begin{pmatrix} -p \\ r \end{pmatrix} = -p\frac{\partial}{\partial r} + r\frac{\partial}{\partial p}.$$

Similarly, show that

$$\frac{\partial}{\partial \rho} = \begin{pmatrix} \frac{r}{\sqrt{r^2+p^2}} \\ \frac{p}{\sqrt{r^2+p^2}} \end{pmatrix} = \frac{1}{\sqrt{r^2+p^2}}\left(r\frac{\partial}{\partial r} + p\frac{\partial}{\partial p}\right).$$

Finally, use the chain rule to check that these equalities make sense for partial differentiation; i.e., show that for any differentiable function f from $\mathbb{R}^2$ to $\mathbb{R}$ we have

$$\frac{\partial f}{\partial \theta} = \rho\left(\cos\theta\frac{\partial f}{\partial p} - \sin\theta\frac{\partial f}{\partial r}\right),$$

and similarly for $\frac{\partial f}{\partial \rho}$.

Derivatives of real-valued functions provide natural examples of one-forms. Typically, multivariable calculus texts (such as Marsden, Tromba and Weinstein [MTW, Section 2.5]) define the gradient of a function f of two variables r and p to be the column vector $\nabla f := (\frac{\partial f}{\partial r}, \frac{\partial f}{\partial p})^T$ and encourage the reader to think of the gradient as a vector field. However, careful consideration of the uses of the gradient should convince the reader that the entries of the gradient morally ought to be arranged in the row vector $df := (\frac{\partial f}{\partial r}, \frac{\partial f}{\partial p})$ because the derivative of a real-valued function at a point $(r_0, p_0)^T$ is truly a linear function from tangent vectors at $(r_0, p_0)^T$ to the real line. We call df the *gradient one-form of the function* f.

For example, one application of the gradient is to the chain rule. Suppose **c** is a parametrized curve in the plane. We might interpret **c** as the trajectory of an ant on a plane and f as temperature, or, more precisely, temperature as a function of position on the plane but independent of time. If both f and **c** are differentiable functions, then

$$(f \circ \mathbf{c})^{'}(t) = \mathbf{c}'(t) \cdot \nabla f(\mathbf{c}(t)), \tag{2.3}$$

i.e., the rate of change of the temperature the ant perceives as a function of time can be calculated from the spatial temperature gradient and the velocity of the ant. If the spacial gradient is known, then to calculate the rate of change of the ant's perceived temperature with respect to time all one really needs is the ant's position and its velocity. In other words, if we know that the ant has velocity vector **v** as it passes through the point $(r, p)^T$, then at that moment the rate of change of the temperature with respect to time is

$$\mathbf{v} \cdot \nabla f(r, p) = v_1 \frac{\partial f}{\partial r} + v_2 \frac{\partial f}{\partial p} = \left(\begin{array}{cc} \frac{\partial f}{\partial r} & \frac{\partial f}{\partial p} \end{array} \right) \left(\begin{array}{c} v_1 \\ v_2 \end{array} \right) = df(\mathbf{v}).$$

So if one believes that matrix multiplication is more natural than the dot product, one must concede that the gradient one-form df is more natural in this context than the vector field ∇f.

Thinking of the derivative of a real-valued function at a point as a covector sheds new light on some concepts of multivariable calculus. The statement that the gradient vector $\nabla f(p, r)$ is perpendicular to the level curves of f near $(p, r)^T$ becomes the statement that the level curves of the gradient covector $df(p, r)$ approximate the level curves of f near $(p, r)^T$. This fits more neatly into the theme of local linearity that pervades all of calculus: the derivative at a point is the best linear approximation for the function near that point. One can use this approach to understand why optimization by the method of Lagrange multipliers works: a local maximum or minimum value of a function f on a level curve of a function g can occur only at a point (p, r) where the level curves of the covector $df(p, r)$ are tangent to the given level curve of g, or where $df(p, r) = 0$. See Figure 2.4. In either case, $df(p, r)$ must be

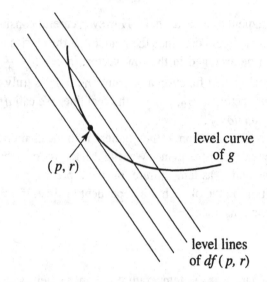

Figure 2.4.   Covector picture for optimization by the method of Lagrange multipliers

a scalar multiple of $dg(p, r)$. Finally, if we think (as physicists often do) of the derivative as the coefficient of the linear term in a Taylor series expansion, we must admit that the derivative at a point of a real-valued function f on $\mathbb{R}^2$ ought to be a linear, real-valued function from $\mathbb{R}^2$ to $\mathbb{R}$, i.e., a covector. Letting that derivative vary from point to point, we obtain the gradient one-form df.

We have already seen one example of a two-form: the area form in Cartesian coordinates. Let us consider another example: the area form in polar coordinates. We use the results of Exercise 26 to calculate the signed area of the parallelogram spanned by $\frac{\partial}{\partial \rho}$ and $\frac{\partial}{\partial \theta}$:

$$A\left(\frac{\partial}{\partial \rho}, \frac{\partial}{\partial \theta}\right) = A\left(\left(\begin{array}{c} \cos\theta \\ \sin\theta \end{array}\right), \left(\begin{array}{c} -\rho\sin\theta \\ \rho\cos\theta \end{array}\right)\right) = \rho.$$

In Section 2.1 we expressed the area form in Cartesian coordinates as a matrix and as a two-form. It is not hard to see that in polar coordinates the matrix is

$$\left(\begin{array}{cc} 0 & \rho \\ -\rho & 0 \end{array}\right)$$

and thus the two-form is $\rho \, d\rho \wedge d\theta$. Notice that the coefficient varies with the coordinates $(\rho, \theta)^T$. So the area form in polar coordinates cannot be considered a constant antisymmetric bilinear form, but it is a two-form.

With these examples in mind, we present a rigorous definition of vector fields, one-forms or two-forms.

Definition 6 *A (differentiable) vector field on* $\mathbb{R}^2$ *is a (differentiable) function from* $\mathbb{R}^2$ *to* $\mathbb{R}^2$. *A (differential) zero-form on* $\mathbb{R}^2$ *is a (differentiable) function from* $\mathbb{R}^2$ *to* $\mathbb{R}$. *A (differential) one-form on* $\mathbb{R}^2$ *is a (differentiable) function from* $\mathbb{R}^2$ *to* $(\mathbb{R}^2)^*$. *A (differential) two-form on* $\mathbb{R}^2$ *is a (differentiable) function from* $\mathbb{R}^2$ *to the space of antisymmetric bilinear forms on* $\mathbb{R}^2$.

In the text we will often omit the words "differentiable" and "differential" when speaking of differentiable vector fields and differential forms.

This definition is a bit too restrictive for our purposes. For instance, polar co-ordinates are defined only for $\rho > 0$, so $\frac{\partial}{\partial \rho}$ is not defined at the origin. Strictly speaking the object $\frac{\partial}{\partial \rho}$ is not a differentiable vector field on $\mathbb{R}^2$. Some readers may wish to handwave this restriction away, thinking of $\frac{\partial}{\partial \rho}$ as a differentiable vector field on $\mathbb{R}^2$ that behaves badly at the origin; this approach will suffice for the purposes of this book. Readers who wish to be rigorous must understand what open sets are; there is a brief introduction with references near the end of Section 3.1. The correct definitions for our purposes are:

Definition 7 *Let U denote an open subset of* $\mathbb{R}^2$. *A differentiable vector field on U is a differentiable function from U to* $\mathbb{R}^2$. *A differential one-form on U is a differentiable function from U to* $(\mathbb{R}^2)^*$. *A differential two-form on U is a differentiable function from U to the space of antisymmetric bilinear forms on* $\mathbb{R}^2$.

So, for example, taking $U = \mathbb{R}^2 \setminus \{(0,0)^T\}$, we say that $\frac{\partial}{\partial \rho}$ is a differentiable vector field on $\mathbb{R}^2 \setminus \{(0,0)^T\}$. Explicitly, we see from Exercise 26 that this differentiable vector field is the function

$$\begin{pmatrix} r \\ p \end{pmatrix} \mapsto \begin{pmatrix} \frac{r}{\sqrt{r^2+p^2}} \\ \frac{p}{\sqrt{r^2+p^2}} \end{pmatrix}.$$

We can take wedge products of differential one-forms to obtain differential two-forms. We start with differential one-forms α and β and differentiable vector fields **v** and **w**. As in Definition 4, we define

$$(\alpha \wedge \beta)(\mathbf{w}, \mathbf{v}) := (\alpha \mathbf{w})(\beta \mathbf{v}) - (\alpha \mathbf{v})(\beta \mathbf{w}).$$

As before, the wedge product itself is antisymmetric in the sense that, for one-forms α and β, we have $\alpha \wedge \beta = -\beta \wedge \alpha$. Note that as a consequence, for any one-form α, we have $\alpha \wedge \alpha = -\alpha \wedge \alpha$, and hence $\alpha \wedge \alpha = 0$.

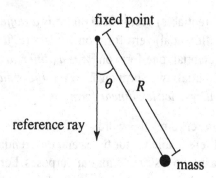

Figure 2.5. Planar pendulum

2.4   The Cylinder and Jacobian Matrices

In this section we consider our first example of a surface that is not itself a Euclidean space: the cylinder. We introduce Jacobian matrices and reinforce the relationship between partial differential operators and geometric vectors.

Consider an infinite cylinder with radius R, centered around the x-axis. Call the cylinder C_R. We can write C_R parametrically:

$$C_R = \{(p, R\cos\theta, R\sin\theta)^T : (\theta, p)^T \in \mathbb{R}^2\}.$$

We can interpret the cylinder as a phase space. Consider a particle whose configuration space is a circle of radius R. For instance, consider any planar pendulum with a bob length of R. See Figure 2.5. Then we can interpret θ as the angle between the bob and a reference ray, and we can interpret p as the momentum of the particle. Specifically, we interpret $|p|$ as the magnitude of the linear momentum of the particle and interpret the sign of p as the direction. A plus sign indicates counterclockwise motion and a minus sign indicates clockwise motion. In Section 4.4 we will analyze two mechanical systems with cylindrical phase spaces.

Let us show that the area form on the cylinder can be expressed as

$$R \, dp \wedge d\theta$$

in differential form notation. The situation here is more complicated than in the planar example (of course, because linear objects such as planes are always the simplest). Recall from multivariable calculus that we can find the surface area of the region on the cylinder that corresponds to a region U in the θp-plane: the answer is $R(\text{area of } U)$. For readers who do not remember how to do this calculation, we provide details. We use a standard technique, following the notation of [MTW, p. 387]. Let us name the function used to parametrize the cylinder C_R: define

$$\Psi : \qquad \mathbb{R}^2 \to \mathbb{R}^3$$

$$\begin{pmatrix} \theta \\ p \end{pmatrix} \mapsto \begin{pmatrix} p \\ R\cos\theta \\ R\sin\theta \end{pmatrix}.$$

To find the integrand for the surface area integral we must calculate

$$\frac{\partial \Psi}{\partial \theta} = \begin{pmatrix} 0 \\ -R\sin\theta \\ R\cos\theta \end{pmatrix}$$

$$\frac{\partial \Psi}{\partial p} = \begin{pmatrix} 1 \\ 0 \\ 0 \end{pmatrix}.$$

Then the surface area integrand, in θp-coordinates, is

$$\left| \frac{\partial \Psi}{\partial \theta} \times \frac{\partial \Psi}{\partial p} \right| = \left| \begin{pmatrix} 0 \\ -R\sin\theta \\ R\cos\theta \end{pmatrix} \times \begin{pmatrix} 1 \\ 0 \\ 0 \end{pmatrix} \right| = \left| \begin{pmatrix} 0 \\ R\cos\theta \\ R\sin\theta \end{pmatrix} \right| = |R| = R.$$

So the surface area in question is

$$\iint_U R\,dp\,d\theta = R(\text{area of } U).$$

Hence for any vectors $\mathbf{v}$ and $\mathbf{w}$ parallel to the θp-plane, the area on the cylinder corresponding to the parallelogram spanned by $\mathbf{v}$ and $\mathbf{w}$ in the θp-plane is $|R\,dp \wedge d\theta(\mathbf{v}, \mathbf{w})|$. So we can take $R\,dp \wedge d\theta$ to be the (signed) area form on the cylinder. The cylinder with the area form $R\,dp \wedge d\theta$ is a symplectic manifold. It is the phase space of a particle constrained to a circle of radius R. The area form expresses the kinematic relationship between the particle's position and its momentum.

Let us try to interpret first-order partial differential operators on the cylinder as vectors, as we did for the plane in Section 2.2. Because θ and p are coordinates on the cylinder, we can differentiate functions on the cylinder with respect to θ or p. For example, consider the function f that calculates the distance from a point in three-space to the origin and define $F := f \circ \Psi$. We remark that the function F has no particular mechanical significance; it is just a *test function* to be plugged into our formalism. Let us differentiate this function with respect to θ and p. Explicitly,

$$\frac{\partial}{\partial p}\left(f(p, R\cos\theta, R\sin\theta)\right) = \frac{\partial}{\partial p}\sqrt{(p^2 + R^2)} = \frac{p}{\sqrt{(p^2 + R^2)}},$$

while

$$\frac{\partial}{\partial \theta}\left(f(p, R\cos\theta, R\sin\theta)\right) = \frac{\partial}{\partial \theta}\sqrt{(p^2 + R^2)} = 0.$$

This makes sense: Going around the cylinder (changing θ while holding p constant) does not change the distance from the origin, while traveling along the axis direction (changing p) does.

Next let us do the same calculation for an arbitrary function F that comes from a function f on three-space, just as our previous F came from the distance function. We will use Jacobian matrices, known also as derivative matrices (see [MTW, p. 129]) and the multivariable chain rule (see [MTW, p. 142]). The Jacobian matrix of a function is the matrix of all possible first partial derivatives with one column for each independent (domain) variable and one row for each dependent (range) variable. For instance, the Jacobian matrix of Ψ at $(\theta, p)^T$ is

$$J_\Psi(\theta, p) = \begin{pmatrix} 0 & 1 \\ -R\sin\theta & 0 \\ R\cos\theta & 0 \end{pmatrix}.$$

In the case of Jacobians we allow typographical concerns to dominate parallelism and write $J_\Psi(\theta, p)$ for $J_\Psi\begin{pmatrix} \theta \\ p \end{pmatrix}$. Likewise, we will allow ourselves to write (θ, p) for a point in the parameter space when we really mean $(\theta, p)^T$. For an arbitrary function $f : \mathbb{R}^3 \to \mathbb{R}$ we have, via the chain rule,

$$J_{f\circ\Psi}(\theta, p) = J_f(\Psi(\theta, p)) \cdot J_\Psi(\theta, p).$$

Equivalently, in explicit matrix form,

$$\begin{pmatrix} \frac{\partial f\circ\Psi}{\partial \theta} & \frac{\partial f\circ\Psi}{\partial p} \end{pmatrix} = \begin{pmatrix} \frac{\partial f}{\partial x} & \frac{\partial f}{\partial y} & \frac{\partial f}{\partial z} \end{pmatrix} \begin{pmatrix} 0 & 1 \\ -R\sin\theta & 0 \\ R\cos\theta & 0 \end{pmatrix}$$

$$= \begin{pmatrix} -R\sin\theta\frac{\partial f}{\partial y} + R\cos\theta\frac{\partial f}{\partial z} & \frac{\partial f}{\partial x} \end{pmatrix}.$$

To see that this makes sense, let us check our distance function, recalling that we evaluate the partial derivatives of f at $(x, y, z) = \Psi(\theta, p) = (p, R\cos\theta, R\sin\theta)$, so our final expressions should be in terms of θ and p:

$$\frac{\partial f}{\partial x} = x(x^2 + y^2 + z^2)^{-\frac{1}{2}} = p(p^2 + R^2)^{-\frac{1}{2}}$$

$$\frac{\partial f}{\partial y} = y(x^2 + y^2 + z^2)^{-\frac{1}{2}} = R\cos\theta(p^2 + R^2)^{-\frac{1}{2}}$$

$$\frac{\partial f}{\partial z} = z(x^2 + y^2 + z^2)^{-\frac{1}{2}} = R\sin\theta(p^2 + R^2)^{-\frac{1}{2}}.$$

So, in our example of the distance function, we find, using the chain rule, that

$$\begin{aligned}
\frac{\partial f \circ \Psi}{\partial \theta} &= -R\sin\theta\frac{\partial f}{\partial y} + R\cos\theta\frac{\partial f}{\partial z} \\
&= -\frac{R\sin\theta(R\cos\theta)}{\sqrt{(p^2 + R^2)}} + \frac{R\cos\theta(R\sin\theta)}{\sqrt{(p^2 + R^2)}} \\
&= 0
\end{aligned}$$

and $\dfrac{\partial f \circ \Psi}{\partial p} = \dfrac{\partial f}{\partial x} = \dfrac{p}{\sqrt{(p^2 + R^2)}},$

which matches what we found by direct calculation.

What is the point? What has this to do with differential operators? The point is that our calculation with an arbitrary f lets us legitimately write

$$\frac{\partial}{\partial p} = \frac{\partial}{\partial x} \quad \text{and} \quad \frac{\partial}{\partial \theta} = -R\sin\theta\frac{\partial}{\partial y} + R\cos\theta\frac{\partial}{\partial z}.$$

We remark that in this example the differential operator $\frac{\partial}{\partial\theta}$ depends on the coordinates. This is quite typical. In general, the coefficients of a vector field can depend on the coordinates. If we interpret (as we did in the plane example)

$$\frac{\partial}{\partial x} = \begin{pmatrix} 1 \\ 0 \\ 0 \end{pmatrix}, \frac{\partial}{\partial y} = \begin{pmatrix} 0 \\ 1 \\ 0 \end{pmatrix}, \frac{\partial}{\partial z} = \begin{pmatrix} 0 \\ 0 \\ 1 \end{pmatrix},$$

then we can think of $\frac{\partial}{\partial\theta}$ and $\frac{\partial}{\partial p}$ as tangent vectors to the cylinder, with $\frac{\partial}{\partial\theta} = (0, -R\sin\theta, R\cos\theta)^T$ and $\frac{\partial}{\partial p} = (1, 0, 0)^T$. Where have we seen these vectors before? They are the vectors $\frac{\partial\Psi}{\partial\theta}$ and $\frac{\partial\Psi}{\partial p}$ we used to calculate surface area.

The reader may also recognize these as the columns of the Jacobian matrix. Jacobian matrices are chock full of geometric significance. This is not surprising once one realizes that they are natural generalizations of the derivative function from single variable calculus and the gradient for scalar-valued functions of several variables.

Exercise 27 *Show that for any real-valued function* $f : \mathbb{R}^n \to \mathbb{R}$ *we have* $J_f = df$ *and if* $n = 1$, *then* $J_f = f'$.

We strongly encourage students to take the time to understand Jacobians fully. Their fundamental importance is often obscured by other material in introductory courses. For an introduction taking advantage of linear algebra, see [HH, Section 1.7].

Exercise 28 *Review your favorite multivariable calculus text to see where the Jacobian matrix and its columns appear, possibly in disguise. At a minimum, recognize their role in surface area integrals, change of variable formulas for multidimensional integrals and Taylor series expansions of functions.*

Exercise 29 *Recall from multivariable calculus that a function* $f : \mathbb{R}^3 \to \mathbb{R}^3$ *can be interpreted as a vector field on* $\mathbb{R}^3$. *Show that the function* $\mathrm{tr} J_f : \mathbb{R}^3 \to \mathbb{R}$ *is equal to the divergence of the vector field.*
Note: the definition of the trace is given in Chapter 0.

We summarize some of the geometric properties of the Jacobian in the context of differentiable functions Ψ parametrizing two-dimensional surfaces, such as the function we used earlier to parametrize the cylinder. Suitable generalizations hold for Jacobians of all differentiable functions. Each column of the Jacobian matrix is tangent to the parametrized surface. For any fixed point (θ, p) in the parameter space, $J_\Psi(\theta, p)$ *pushes* vectors tangent to the parameter space (at (θ, p)) *forward* into vectors tangent to the parametrized object (at $\Psi(\theta, p)$). See Figure 2.6. If the Jacobian matrix has full rank at some point (that is, if the two columns of $J_\Psi(\theta, p)$ are not parallel), then the columns of the Jacobian span the tangent plane to the surface. Finally, we can use the Jacobian to find the best first-order approximation to the function Ψ: for $(\tilde{\theta}, \tilde{p})$ sufficiently close to a fixed point (θ, p) we have

$$\Psi(\tilde{\theta}, \tilde{p}) \approx \Psi(\theta, p) + J_\Psi(\theta, p) \begin{pmatrix} \tilde{\theta} - \theta \\ \tilde{p} - p \end{pmatrix}.$$

Note that $J_\Psi(\theta, p)$ is a constant 3×2 matrix, so multiplying it on the right by $(\tilde{\theta} - \theta, \tilde{p} - p)^T$ yields a column three-vector that is a first-degree polynomial in $\tilde{\theta}$ and $\tilde{p}$. If the Jacobian has full rank, then the graph of $\Psi(\theta, p) + J_\Psi(\theta, p)(\tilde{\theta} - \theta, \tilde{p} - p)^T$ is the tangent plane to the surface at the point $\Psi(\theta, p)$. To put it more philosophically, the Jacobian matrix is the key to local linearity.

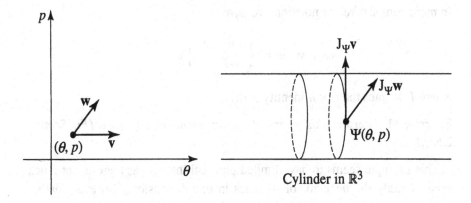

Figure 2.6. Push forwards

Exercise 30 *Using the polar coordinates (ρ, θ) on the plane (see Section 2.3) find $\frac{\partial}{\partial \rho}$ and $\frac{\partial}{\partial \theta}$ in terms of $\frac{\partial}{\partial p}$ and $\frac{\partial}{\partial r}$ by reinterpreting all four differential operators as column vectors and using the Jacobian of the function that parametrizes the plane by polar coordinates. Compare with Exercise 26.*

To reiterate the main conceptual point of this section: sometimes it's helpful to think of $\frac{\partial}{\partial \theta}$ and $\frac{\partial}{\partial p}$ as differential operators; sometimes it's helpful to think of them as vectors, that is, geometric objects determined by their length and direction. *They are both.*

2.5   More Examples of Phase Spaces

One extremely important example is the phase space of n particles moving in $\mathbb{R}$ with position-momentum pairs $(r_1, p_1)^T \dots (r_n, p_n)^T$. Consider $\mathbb{R}^{2n}$ with the *canonical* symplectic form

$$\omega := dp_1 \wedge dr_1 + dp_2 \wedge dr_2 + \cdots + dp_n \wedge dr_n, \qquad (2.4)$$

where we have chosen coordinates on $\mathbb{R}^{2n}$:

$$(r_1, \dots, r_n, p_1 \dots, p_n).$$

In other words, we have

$$\omega \left(\sum_{i=1}^{n} (v_i \frac{\partial}{\partial r_i} + v_{i+n} \frac{\partial}{\partial p_i}), \sum_{i=1}^{n} (w_i \frac{\partial}{\partial r_i} + w_{i+n} \frac{\partial}{\partial p_i}) \right) := \sum_{i=1}^{n} (v_i w_{i+n} - v_{i+n} w_i).$$

In more compact vector notation, we have

$$\omega(\mathbf{v}, \mathbf{w}) = v^T \begin{pmatrix} 0 & I \\ -I & 0 \end{pmatrix} w,$$

where I denotes the $n \times n$ identity matrix.

Exercise 31 *Interpret this symplectic form geometrically. (See [Fr, Section 2.5c].)*

This example seems to have limited physical interest, as there is not much need to study the behavior of particles in one dimension. However, mathematically speaking, this is the single most important example of phase space because of *Darboux's Theorem*, which states that around any point m in any $2n$-dimensional phase space it is possible to find coordinates (called *canonical coordinates*) such that the symplectic form takes the form of Equation 2.4 close to the point m. For a precise statement of this theorem, see [AM, Theorem 3.2.2] or [Ar89, p. 230].

Setting $n = 6$ and making a modification to account for collisions yields the phase space for the two-body problem: the first three coordinates give the position of the first particle; the next three coordinates give the position of the second particle; the next three give the momentum of the first particle and the final three give the momentum of the second particle. To avoid collisions we must rule out the points in $\mathbb{R}^{12}$ corresponding to equal positions for the two particles. That is, the actual phase space is

$$\{(r_1, \ldots, r_6, p_1, \ldots, p_6) \in \mathbb{R}^{12} : (r_1, r_2, r_3) \neq (r_4, r_5, r_6)\}$$

with the symplectic form given in Equation 2.4. Because Kepler's laws concern planets that never collide with the sun, for the purposes of this book we can forget about collisions and think of the phase space of the two-body problem as simply $\mathbb{R}^{12}$. This modification will complicate our lives only once, in Section 8.4, where we will insist on avoiding collisions.

Another example is the phase space of a charged particle moving in $\mathbb{R}^3$ under the influence of a constant magnetic field. The underlying manifold is $\mathbb{R}^6$, but the symplectic form is different. If we let $(B_x, B_y, B_z)^T$ denote the magnetic field and let q denote the charge of the particle, then we can encode the kinetics of the particle in

$$\omega := dp_x \wedge dr_x + dp_y \wedge dr_y + dp_z \wedge dr_z$$
$$+ q(B_x dr_y \wedge dr_z + B_y dr_z \wedge dr_x + B_z dr_x \wedge dr_y).$$

We will see (in Section 4.3) that this symplectic form encodes correctly the effects of the magnetic force.

Yet another example is the phase space of the spherical pendulum, which is a particle constrained to move on a sphere. Because the momentum vector must be geometrically tangent to the sphere and hence perpendicular to the position vector, we find that the set of all possible position-momentum pairs is

$$T^*S^2 := \{(\mathbf{r}, \mathbf{p}) \in \mathbb{R}^3 \times (\mathbb{R}^3)^* : |\mathbf{r}| = 1 \text{ and } \mathbf{pr} = 0\}.$$

The symplectic form is

$$\omega := dp_x \wedge dr_x + dp_y \wedge dr_y + dp_z \wedge dr_z.$$

We choose the symbol "T^*S^2" because it is the simplest common name for this space. (Sophisticated readers may recognize this as the cotangent bundle of the sphere S^2; such readers should also note that all of the phase spaces we have discussed are cotangent bundles: our $\mathbb{R}^{2n}$ is $T^*\mathbb{R}^n$ and the cylinder is T^*S^1. In general, the phase space of a mechanical system is the cotangent bundle of its configuration space.)

Exercise 32 *Use the implicit function theorem to show that T^*S^2 can be parametrized differentiably near any point by four real parameters.*

2.6   Pullback Calculations

Our final task in this chapter is to practice changing coordinates and finding expressions for old forms in the new coordinates. This is known as *pulling back* forms. In Section 2.3 we introduced differential one-forms and differential two-forms; in this context it is natural to think of functions as differential zero-forms. We will practice pulling forms back from Cartesian (r, p) coordinates to polar (ρ, θ) coordinates.

First, we call attention to the transition function itself. To indicate the relevance of this example to Definition 9 in Section 3.1, we use the capital Greek letter gamma (Γ):

$$\Gamma : \mathbb{R}^+ \times \mathbb{R} \to \mathbb{R}^2 \setminus \{(0, 0)^T\}$$
$$\begin{pmatrix} \rho \\ \theta \end{pmatrix} \mapsto \begin{pmatrix} \rho \cos \theta \\ \rho \sin \theta \end{pmatrix}. \tag{2.5}$$

This function goes from polar coordinates to Cartesian coordinates; the "back" in "pullback" refers to going from Cartesian to polar coordinates. We will find the Jacobian matrix of Γ useful, so we calculate it here:

$$J_\Gamma(\rho, \theta) = \begin{pmatrix} \cos\theta & -\rho\sin\theta \\ \sin\theta & \rho\cos\theta \end{pmatrix}.$$

Pulling zero-forms back is easy. Given a function f in Cartesian coordinates, we can simply think of it as a function in polar coordinates. In physics applications, our function f will usually be naturally defined in some coordinate-free way. For instance, we might think of distance from the origin or of a given temperature distribution which we can describe in several different coordinate systems. In such a situation it makes sense to use the same letter (such as ρ or T) for the function no matter what coordinates we choose to work in. However, to be mathematically rigorous we cannot use the same name for two functions that perform different computations. To calculate the distance from the origin in polar coordinates, we simply take the value of the first argument, while in Cartesian coordinates we take the square root of the sum of the squares of the two arguments. These are two quite different functions, so it is mathematically more honest to use notation that distinguishes them. Naming the transition function Γ explicitly gives us a natural way to be mathematically honest while retaining in our notation the relationship of two expressions for the same quantity in different coordinates. So for any function f in Cartesian coordinates, the composition $f \circ \Gamma$ is the corresponding function in polar coordinates. This composition is called the *pullback of f along* Γ. We will use the notation $\Gamma^* f$ for $f \circ \Gamma$.

Exercise 33 *Let f be the distance function in Cartesian coordinates. Write a formula for f and calculate $\Gamma^* f$ explicitly. Your answer should be the function* $(\rho, \theta)^T \mapsto \rho$.

Next we pull back one-forms. Again, we can follow our calculational gut instincts. For example, to pull the one-form dr back to polar coordinates, we can think of r as a function on the plane and recall from Section 2.3 that dr is the one-form version of the gradient. So in polar coordinates we have

$$dr = d(\rho\cos\theta) = \begin{pmatrix} \frac{\partial}{\partial\rho}(\rho\cos\theta) & \frac{\partial}{\partial\theta}(\rho\cos\theta) \end{pmatrix} = \begin{pmatrix} \cos\theta & -\rho\sin\theta \end{pmatrix}$$
$$= \cos\theta \, d\rho - \rho\sin\theta \, d\theta.$$

It is useful to notice that we get the same result by applying differentiation rules (such as the product and chain rules) to d:

$$dr = d(\rho\cos\theta) = (d\rho)\cos\theta + \rho(d\cos\theta)$$
$$= \cos\theta \, d\rho - \rho\sin\theta \, d\theta.$$

Exercise 34 *Show that*

$$dp = \sin\theta \, d\rho + \rho \cos\theta \, d\theta.$$

To give a mathematically rigorous definition of the pullback $\Gamma^*\alpha$ of a one-form α along a function Γ, we think of α as a row vector and multiply it by the matrix J_Γ.

Exercise 35 *Using the fact that $dr = (1, 0)$ in Cartesian coordinates, use the Jacobian J_Γ to calculate once again that $dr = \cos\theta \, d\rho - \rho \sin\theta \, d\theta$.*

Pulling back wedge products of one-forms is easy because, as a mathematician might say, pullbacks respect the wedge. In other words, if α and β are one-forms in the variables r and p, then

$$\Gamma^* (\alpha \wedge \beta) = (\Gamma^*\alpha) \wedge (\Gamma^*\beta).$$

We can apply this technology to the area form in Cartesian coordinates to obtain the area form in polar coordinates. We know from Section 2.2 that the area form in Cartesian coordinates is $dr \wedge dp$.

$$
\begin{aligned}
\Gamma^* (dr \wedge dp) &= (\Gamma^*dr) \wedge (\Gamma^*dp) \\
&= (\cos\theta d\rho - \rho \sin\theta d\theta) \wedge (\sin\theta d\rho + \rho \cos\theta d\theta) \\
&= \cos\theta \sin\theta d\rho \wedge d\rho - \rho \sin^2\theta d\theta \wedge d\rho + \rho \cos^2\theta d\rho \wedge d\theta \\
&\quad - \rho^2 \sin\theta \cos\theta d\theta \wedge d\theta.
\end{aligned}
$$

By the antisymmetry of the wedge product we know that $d\rho \wedge d\theta = -d\theta \wedge d\rho$ and $d\rho \wedge d\rho = d\theta \wedge d\theta = 0$. So we can conclude that

$$\Gamma^* (dr \wedge dp) = \rho \, d\rho \wedge d\theta.$$

Because the relationship of polar coordinates to Cartesian coordinates is clear, we can suppress the Γ^* and write $dr \wedge dp = \rho \, d\rho \wedge d\theta$. Of course, this result is not news to us; we calculated it in Section 2.3. Our pullback technology has given us an alternative derivation.

It should be clear that the procedure we used for polar coordinates works for any change of coordinates. One can always reexpress a form in new coordinates by writing the old coordinates as an explicit function Γ and calculating the pullback. Equivalently, one can plug the formulas for the old coordinates into the forms and calculate, following the calculational rules for linear combinations and wedges, with d acting as a derivative. This second notation is simpler because it avoids introducing Γ, and will work well in any situation where the variables all have clear physical meaning.

Note that, except for the wedge itself, this expression for the area form in polar coordinates is the integration factor for functions on the plane in polar coordinates. This is an indication of an important role that forms play in differential geometry: forms were born to be integrated. One-forms get integrated over curves and two-forms get integrated over two-dimensional surfaces. Although the wedge may make the formula look different from the line integral and surface integral formulas learned in a standard undergraduate vector calculus class, it is just a convenient way to keep track of minus signs. For the two body problem, we do not need to integrate our forms, so we will not provide any more detail on this aspect of them. However, we do use the idea of integrating forms in the optional Chapter 3 when we define closed forms. Interested readers should consult [Fr, Chapter 3], [HH, Chapter 6] or [Mu, Chapter 7].

Pullbacks also respect linear operations. The pullback of the sum of two differential forms is the sum of the pullback; likewise, the pullback of a scalar multiple of a differential form is the same scalar multiple of the pullback.

Unlike differential forms, vector fields do not behave naturally under pullbacks. *Pushing* them *forward* is sometimes possible, but not always. For instance, for the function Γ defined by Formula 2.5 the vector field $\theta \frac{\partial}{\partial \theta}$ is well-defined on the domain $\mathbb{R}^+ \times \mathbb{R}$ but it makes no sense on the range because θ is defined only up to multiples of 2π. However, in some situations it is possible to push vector fields forward. A sufficient technical condition to impose is that the function Γ should be injective and surjective. For instance, if we restrict the domain in Formula 2.5 to $\mathbb{R}^+ \times (0, 2\pi)$, then we can push $\theta \frac{\partial}{\partial \theta}$ forward without ambiguity. From Exercise 30 we know that $\frac{\partial}{\partial \rho} = \cos\theta \frac{\partial}{\partial r} + \sin\theta \frac{\partial}{\partial p}$ and $\frac{\partial}{\partial \theta} = -\rho \sin\theta \frac{\partial}{\partial r} + \rho \cos\theta \frac{\partial}{\partial p}$. This matches the rigorous definition of the *pushforward*: the pushforward of a vector field in the domain of Γ is obtained by multiplying the matrix J_Γ by the column vector of the vector field:

$$\begin{pmatrix} \cos\theta & -\rho\sin\theta \\ \sin\theta & \rho\cos\theta \end{pmatrix} \begin{pmatrix} 1 \\ 0 \end{pmatrix} = \begin{pmatrix} \cos\theta \\ \sin\theta \end{pmatrix}$$

and

$$\begin{pmatrix} \cos\theta & -\rho\sin\theta \\ \sin\theta & \rho\cos\theta \end{pmatrix} \begin{pmatrix} 0 \\ 1 \end{pmatrix} = \begin{pmatrix} \rho\sin\theta \\ -\rho\cos\theta \end{pmatrix}.$$

The pushforward of a vector field is a precise way of reexpressing in the old coordinates a vector field given in terms of the new coordinates.

Exercise 36 *Define new coordinates by*

$$\begin{pmatrix} r \\ p \end{pmatrix} = M \begin{pmatrix} x \\ y \end{pmatrix},$$

where M is a constant 2×2 invertible matrix.

1. *Use the pushforward to find $\frac{\partial}{\partial r}$ and $\frac{\partial}{\partial p}$ in terms of $\frac{\partial}{\partial x}$ and $\frac{\partial}{\partial y}$. Then find $\frac{\partial}{\partial x}$ and $\frac{\partial}{\partial y}$ in terms of $\frac{\partial}{\partial r}$ and $\frac{\partial}{\partial p}$.*

2. *Find dx and dy in terms of dr and dp from the relations $dx(\frac{\partial}{\partial x}) = 1$, $dx(\frac{\partial}{\partial y}) = 0$, etc.*

3. *Find dx and dy in terms of dr and dp by pulling back.*

Express all answers (and, if you can, all calculations) in terms of the matrix M, not in terms of its entries.

3

Bridge to Differential Geometry

Except for a few exercises (marked "optional"), later chapters do not depend crucially on the contents of this chapter. The reader should feel free to skip directly to Chapter 4.

One of the goals of this book is to motivate students to learn *differential geometry*, the study of manifolds and the geometrically natural objects defined on them (such as vector fields and differential forms). Differential geometry is basic not only to classical mechanics but to many other fields of physics. Its importance in general relativity is well-known, but it also applies to any field of inquiry in which quantities change smoothly most of the time, including statistical mechanics and economics. Regardless of external applications, differential geometry is universally appreciated by mathematicians for its beauty and wide-ranging utility within mathematics.

Many colleges and universities offer courses at the advanced undergraduate or graduate level, and there are many excellent books available, some of which are listed in the Recommended Reading. This chapter attempts to bridge the gap between the abstract, rigorous development of the concepts of differential geometry and the informal limited introduction provided in Chapter 2.

3.1 Manifolds

The most natural and fundamental concept of differential geometry (and, in fact, of much of modern mathematics) is that of a *manifold*. At the same time, the rigorous definition of the term is, by an order of magnitude, one of the longest, most complex and often most off-putting of all the definitions a fledgling mathematician encounters. Dear reader, we will try to be gentle, and we hope you will trust us that the end is worthy of the means. We will start with an informal version of the definition, consider two examples, and then show how to turn our ideas into a formal definition.

Definition 8 (informal) *A set M is an n-dimensional manifold if it satisfies both:*

1. *Near each point $p \in M$ there is a good system of n coordinates, and*

2. *Coordinate systems are compatible with one another.*

When we give the formal definition (Definition 9) we will make precise what "good" and "compatible" mean. Intuitively, good coordinates should be well-defined functions on the part of the manifold they parametrize; good coordinates should also be appropriate independent variables for differentiation. "Compatibility" means that on any overlap the change-of-coordinates function should be infinitely differentiable.

For example, the circle is a one-dimensional manifold. Except for one fixable glitch, the usual angle variable is a good coordinate. Let us do this example in detail. We think of the circle as the set S^1 of points $\begin{pmatrix} x \\ y \end{pmatrix}$ in $\mathbb{R}^2$ satisfying $x^2 + y^2 = 1$. We can parametrize the circle (several times over) by the function

$$\gamma : \quad \mathbb{R} \to S^1$$
$$t \mapsto \begin{pmatrix} \cos t \\ \sin t \end{pmatrix}.$$

See Figure 3.1.

We cannot use this t as a coordinate because t is not a well-defined function on the circle: for each point $\begin{pmatrix} x \\ y \end{pmatrix}$ on the circle (such as $\begin{pmatrix} 1 \\ 0 \end{pmatrix}$) there are several corresponding values of t (such as $0, \pm 2\pi, \pm 4\pi, \ldots$). We cannot fix

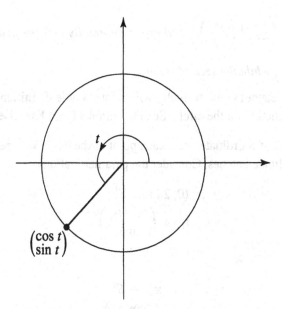

Figure 3.1.   The circle and its usual paramterization

the problem simply by restricting. If we try the parametrization

$$\tilde{\gamma}: \quad [0, 2\pi) \to S^1 \tag{3.1}$$

$$t \mapsto \begin{pmatrix} \cos t \\ \sin t \end{pmatrix}, \tag{3.2}$$

then t is well-defined as a function on the circle but there is a problem with differentiation at the point $\begin{pmatrix} 1 \\ 0 \end{pmatrix}$. Because the point $t = 0$ lies on the boundary of the interval $[0, 2\pi)$, we cannot naively take a derivative with respect to t there. Although there are ways to get around the problem in this case by considering "one-sided" derivatives, we want to keep our constructions simple so that we can generalize them. There is no satisfying way to encode the information that points on the circle with t close to 2π are near the point with $t = 0$. So the parametrization $\tilde{\gamma}$ is not good enough. And we cannot fix it by extending the interval of definition because that would make t multivalued at some points.

Exercise 37 *Consider the functions* $[0, 2\pi) \to \mathbb{R}$, $t \mapsto \sin t$ *and* $[0, 2\pi) \to \mathbb{R}$, $t \mapsto \sin \frac{t}{2}$. *First check that both functions are differentiable on* $(0, 2\pi)$ *and have one-sided derivatives at 0. Next, consider the corresponding functions on the circle* $S^1 \to \mathbb{R}$, $\begin{pmatrix} x \\ y \end{pmatrix} \mapsto \sin \left(\tilde{\gamma}^{-1} \begin{pmatrix} x \\ y \end{pmatrix} \right)$ *and* $S^1 \to \mathbb{R}$,

$\begin{pmatrix} x \\ y \end{pmatrix} \mapsto \sin\left(\frac{1}{2}\tilde{\gamma}^{-1}\begin{pmatrix} x \\ y \end{pmatrix}\right)$ *and argue informally that the first is differen-*

tiable at $\begin{pmatrix} 1 \\ 0 \end{pmatrix}$ *while the second is not.*

Note: that you cannot argue formally without a suitable definition of differentiability of a function on the circle. See Definition 11 and Exercise 49.

In order to have coordinates near each point of the circle we need to use two different coordinate patches. Consider the parametrizations

$$\gamma_1 : \quad (0, 2\pi) \to S^1$$
$$t_1 \mapsto \begin{pmatrix} \cos t_1 \\ \sin t_1 \end{pmatrix} \tag{3.3}$$

and

$$\gamma_2 : \quad (-\pi, \pi) \to S^1$$
$$t_2 \mapsto \begin{pmatrix} \cos t_2 \\ \sin t_2 \end{pmatrix}. \tag{3.4}$$

Notice that each point on the circle corresponds to at most one value of t_1 and one value of t_2. Because t_1 and t_2 both range over open intervals, each is a good coordinate. We must also check the overlap condition. At every point $\begin{pmatrix} x \\ y \end{pmatrix}$ on the circle with $y \neq 0$, both t_1 and t_2 are well-defined. On the top half of the circle $t_1 = t_2$, so the change-of-coordinates in either direction is just the identity function, which is infinitely differentiable. On the bottom half, $t_1 = t_2 + 2\pi$, so the change-of-coordinates functions are $t_2 \mapsto t_2 + 2\pi$, and $t_1 \mapsto t_1 - 2\pi$, both of which are infinitely differentiable (see Figure 3.2). So the circle S^1 is a one-manifold.

Another important example is the sphere S^2. Let us show that it is a two-manifold. We can parametrize the unit sphere (many times over) with the usual spherical coordinates

$$\gamma : \qquad \mathbb{R}^2 \to S^2$$
$$\begin{pmatrix} \theta \\ \phi \end{pmatrix} \mapsto \begin{pmatrix} \sin\theta \cos\phi \\ \sin\theta \sin\phi \\ \cos\theta \end{pmatrix}.$$

Here we have used the physicists' convention: ϕ is the longitude (curves of constant ϕ correspond to meridians from one pole to the other) and θ is the colatitude (curves of constant θ correspond to circles parallel to the equatorial

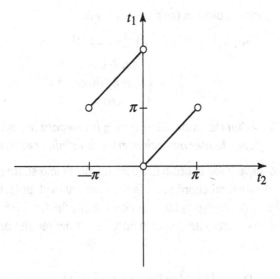

Figure 3.2. A transition function for the circle manifold

circle). The mathematicians' convention reverses the roles of ϕ and θ and has the virtue that the θ of spherical coordinates corresponds to the θ of cylindrical coordinates.

Exercise 38 *Use spherical coordinates (with ϕ representing the standard longitude and with $\theta = 0$ at the north pole) to describe the north and south poles, the Greenwich meridian, the international date line, the equator, the tropic of capricorn and your current location.*

As we saw in the circle example, a good parametrization, while helpful, is not necessarily the whole story. To build a good set of coordinate systems from a parametrization we must fix redundancies. As before, we can get one good coordinate system by restricting the domain of the parametrization:

$$\gamma_1: \qquad (0, \pi) \times (-\pi, \pi) \to S^2$$

$$\begin{pmatrix} \theta_1 \\ \phi_1 \end{pmatrix} \mapsto \begin{pmatrix} \sin\theta_1 \cos\phi_1 \\ \sin\theta_1 \sin\phi_1 \\ \cos\theta_1 \end{pmatrix}.$$

Exercise 39 *Check that θ_1 and ϕ_1 are well-defined functions on most of the sphere.*

This coordinate system omits only the north pole ($\theta = 0$), the south pole ($\theta = \pi$) and the international date line ($\phi = \pi$). We can take care of the international date line by the trick we used in the circle example:

$$\gamma_2 : \qquad (0, \pi) \times (0, 2\pi) \to S^2$$

$$\begin{pmatrix} \theta_2 \\ \phi_2 \end{pmatrix} \mapsto \begin{pmatrix} \sin \theta_2 \cos \phi_2 \\ \sin \theta_2 \sin \phi_2 \\ \cos \theta_2 \end{pmatrix}.$$

Exercise 40 *Check that this coordinate system is compatible with the previous one: the change of coordinates on each overlap is infinitely differentiable.*

We still do not have good coordinates near the north and south poles. Notice that the failure of spherical coordinates is dramatic: at each pole, a whole continuum of values of ϕ correspond to one point. Still, the fix is easy. Near each pole we can use xy-coordinates. Specifically, let D denote the open unit disk in the xy-plane, that is,

$$D := \{ \begin{pmatrix} x \\ y \end{pmatrix} \in \mathbb{R}^2 : x^2 + y^2 < 1 \}$$

and define coordinate systems

$$\gamma_3 : \qquad D \to S^2$$

$$\begin{pmatrix} x_1 \\ y_1 \end{pmatrix} \mapsto \begin{pmatrix} x_1 \\ y_1 \\ \sqrt{1 - x_1^2 - y_1^2} \end{pmatrix}$$

near the north pole and

$$\gamma_4 : \quad D \to S^2$$

$$\begin{pmatrix} x_2 \\ y_2 \end{pmatrix} \mapsto \begin{pmatrix} x_2 \\ y_2 \\ -\sqrt{1 - x_2^2 - y_2^2} \end{pmatrix}$$

near the south pole.

Exercise 41 *Check that the coordinate system defined by γ_3 is compatible with the coordinate system defined by γ_1.*

The four coordinate systems, taken together, suffice for the proof that the sphere is a two-manifold.

Exercise 42 *Use the four given coordinate systems to show (informally) that the sphere is a manifold.*
Note: if you've done the previous three exercises, you've done all of the hard work already.

Exercise 43 *Find two coordinate systems that together suffice to show that the sphere is a manifold.*

Notice that the domains of definition for $\gamma_1, \gamma_2, \gamma_3$ and γ_4 are *open* sets in $\mathbb{R}^2$, i.e., for each element $\begin{pmatrix} x \\ y \end{pmatrix}$ of each domain there is a strictly positive number r such that the ball of radius r around $\begin{pmatrix} x \\ y \end{pmatrix}$

$$B\left(r, \begin{pmatrix} x \\ y \end{pmatrix}\right) := \{ \begin{pmatrix} u \\ v \end{pmatrix} \in \mathbb{R}^2 : (u - x)^2 + (v - y)^2 < r \}$$

lies entirely in the given domain.

Intuitively, every point of an open set lies fully inside the set. So if the domain of the parametrization is open, the coordinates will lend themselves well to differentiation. The problem that arose with the parametrization $\tilde{\gamma}$ in the analysis of the circle was possible only because the domain $[0, 2\pi)$ was not an open set. (For an introduction to open sets, see [Ba, Section 9] or [MH, Section 2.1].)

Exercise 44 *Prove that each of the three sets $(0, \pi) \times (-\pi, \pi)$, $(0, \pi) \times (0, 2\pi)$ and D is open in $\mathbb{R}^2$.*

In the formal definition we replace the vague notion of a "good system of n coordinates" with the precise concept of an injective function whose domain is an open subset of $\mathbb{R}^n$. Our definition is very close to that of Warner [Wa, Definition 1.1.4], whose clear, rigorous treatment we highly recommend.

Definition 9 (Formal) *An n-dimensional manifold is a set M and a family $\{\gamma_1, \gamma_2, \dots\}$ of injective functions from domains in $\mathbb{R}^n$ into M such that*

1. *For each i the domain U_i of the function γ_i is an open subset of $\mathbb{R}^n$. The domains U_i and the functions γ_i are called* coordinate patches.

2. *The union of the coordinate patches cover M. That is, $\gamma_1(U_1) \cup \gamma_2(U_2) \cup \dots = M$.*

3. *The coordinate systems are compatible on overlaps. That is, if we consider the overlaps $W_{ij} := \gamma_i(U_i) \cap \gamma_j(U_j)$ then the sets $\gamma_i^{-1}(W_{ij})$ and $\gamma_j^{-1}(W_{ij})$ must be open in $\mathbb{R}^n$ and the transition functions*

$$\Gamma_{ji} := \gamma_j^{-1} \circ \gamma_i : \gamma_i^{-1}(W_{ij}) \to \gamma_j^{-1}(W_{ij})$$

must be infinitely differentiable.

Let us review our example of the circle S^1 in light of this definition. The domains in $\mathbb{R}$ are $U_1 = (0, 2\pi)$ and $U_2 = (-\pi, \pi)$; these are open, satisfying condition 1 of the definition. The corresponding coordinate patches on the circle itself are $\gamma_1(U_1) = S^1 \setminus \left\{ \begin{pmatrix} 1 \\ 0 \end{pmatrix} \right\}$ and $\gamma_2(U_2) = S^1 \setminus \left\{ \begin{pmatrix} -1 \\ 0 \end{pmatrix} \right\}$. The union of these two sets is S^1 itself, so condition 2 is satisfied. Finally, we must check condition 3 for the overlap $W_{21} = S^1 \setminus \left\{ \begin{pmatrix} 1 \\ 0 \end{pmatrix}, \begin{pmatrix} -1 \\ 0 \end{pmatrix} \right\}$. First we note that $\gamma_1^{-1}(W_{21}) = (0, \pi) \cup (\pi, 2\pi)$, which is open in $\mathbb{R}$. Similarly, $\gamma_1^{-1}(W_{21})$ is open in $\mathbb{R}$. Next, the transition function $\Gamma_{12} : (-\pi, 0) \cup (0, \pi) \to (0, \pi) \cup (\pi, 2\pi)$ can be written explicitly as

$$\Gamma_{12} : t_2 \mapsto \begin{cases} t_2 + 2\pi & \text{if } t_2 \in (-\pi, 0) \\ t_2 & \text{if } t_2 \in (0, \pi) \end{cases}.$$

See Figure 3.2 for the graph of this function, which is infinitely differentiable at every point where it is defined. (Note that the jump at $t_2 = 0$ is not a problem because 0 is not in the domain of the function.) Similarly, $\gamma_2^{-1} \circ \gamma_1$ is infinitely differentiable on its domain. So the circle S^1 satisfies our formal definition of a one-manifold.

Exercise 45 *Show formally that for any natural number k the Euclidean space $\mathbb{R}^k$ is a k-manifold.*

Exercise 46 *Show formally that the sphere is a two-manifold.*

Exercise 47 *Show that T^*S^2, defined in Section 2.5, is a four-manifold. The implicit function theorem will be helpful. (Compare with Exercise 32.)*

The sophisticated reader may have noticed that we have swept some subtleties under the rug. First, our definition allows various strange-looking spaces to qualify as manifolds. To rule out these pathologies we would have to introduce some new concepts, such as *Hausdorff spaces* and *second-countability*. Second, we have implicitly required the number of coordinate patches to be *countable*, i.e., in one-to-one correspondence with the natural numbers. This is not necessary or even desirable mathematically. However, it suffices for the purposes of this book and it makes the notation easier. The third disadvantage of our definition is that it allows us to distinguish between, say, two circles positioned differently in the plane or, even worse, two different sets of parametrizations of the same circle. Intuitively, all circles of the same radius are essentially the same; one might go so far as to argue that all circles, regardless of radius, are essentially the same. A complete treatment of manifolds should introduce the notion of *diffeomorphisms* between manifolds. While these concepts are crucial to a deep mathematical understanding of differential geometry, they are rarely relevant in classical mechanics, which is the main subject of this book. Warner's introduction to manifolds [Wa] deals with these issues openly. For detailed discussion of the underlying topological concepts, the reader might consult Kelley [K].

Exercise 48 *Show that if M is an m-dimensional manifold and N is an n-dimensional manifold, then the Cartesian product M × N has the structure of an (m + n)-dimensional manifold.*

3.2 Differentiable Functions

The first big payoff from the definition of a manifold is that we can define the notion of a differentiable function from one manifold to another. To test differentiability of a function, we express the function in terms of coordinates on the range and domain and test differentiability of the coordinate expressions. We will start with an informal definition (Definition 10) and do a few examples in detail before giving a formal definition (Definition 11).

Definition 10 (informal) *A function f from a manifold M to a manifold N is infinitely differentiable if it is infinitely differentiable in all coordinate patches.*

Our first example is a function from the circle to the real line. We can ask whether the function

$$f: \quad S^1 \to \mathbb{R}$$
$$\begin{pmatrix} x \\ y \end{pmatrix} \mapsto y$$

is infinitely differentiable. The essence of the calculation is to write this function in each coordinate patch. Here we have two coordinate patches on the domain (see equations 3.3 and 3.4) and one coordinate patch on the range (since the identity function suffices to prove that $\mathbb{R}$ is a one-manifold). Writing f in t_1-coordinates we have

$$f \circ \gamma_1: \quad (0, 2\pi) \to \mathbb{R}$$
$$t_1 \mapsto \sin t_1,$$

while in t_2-coordinates we have

$$f \circ \gamma_2 : \ (-\pi, \pi) \to \mathbb{R}$$
$$t_2 \mapsto \sin t_2.$$

Because these functions are both infinitely differentiable, f is infinitely differentiable. This function f is essentially the first function defined in Exercise 37.

Exercise 49 *Show that the second function defined in Exercise 37 is not infinitely differentiable.*

Next let us do an example where the function takes values in a nontrivial manifold. We will define a function from the unit circle onto the 40th north parallel on the unit sphere. (On the earth, the 40th north parallel passes through Salt Lake City, Philadelphia, Madrid, Mt. Olympus, Ankara and Beijing.) Specifically, consider

$$g: \quad S^1 \quad \to \quad S^2$$
$$\begin{pmatrix} x \\ y \end{pmatrix} \mapsto \begin{pmatrix} x \sin \frac{5\pi}{18} \\ y \sin \frac{5\pi}{18} \\ \cos \frac{5\pi}{18} \end{pmatrix}.$$

Since we have two coordinate patches on S^1 and four on S^2, we have $2 \times 4 = 8$ coordinate versions of the function g. Fortunately we do not have to check all eight explicitly. Two are completely irrelevant – $x_2 y_2$-coordinates as a function of t_1 or of t_2 – because $\cos \frac{5\pi}{18} > 0$ and so no value of g lies in the portion of the sphere covered by $x_2 y_2$-coordinates. Of the six remaining ways to write g in coordinates we can get away with checking only two: $x_1 y_1$-coordinates as a function of t_1 and $x_1 y_1$-coordinates as a function of t_2. Because these two coordinate versions describe g completely and because the coordinate systems are compatible, differentiability in other coordinates will follow from differentiability in these. We will explain this idea more fully in the remark following Definition 11.

Writing $x_1 y_1$-coordinates as a function of t_1 using g, we have

$$(0, 2\pi) \quad \to \quad D$$
$$t_1 \mapsto \begin{pmatrix} \cos t_1 \sin \frac{5\pi}{18} \\ \sin t_1 \sin \frac{5\pi}{18} \end{pmatrix},$$

which is infinitely differentiable. Likewise,

$$(-\pi, \pi) \quad \to \quad D$$

$$t_2 \quad \mapsto \quad \begin{pmatrix} \cos t_2 \sin \frac{5\pi}{18} \\ \sin t_2 \sin \frac{5\pi}{18} \end{pmatrix}$$

is infinitely differentiable. We conclude that the function $g : S^1 \to S^2$ is infinitely differentiable.

With these examples under our belt let us consider the formal definition.

Definition 11 (Formal) *Given a manifold M, a manifold N and a function $f : M \to N$, we say that f is infinitely differentiable if for every coordinate system $\gamma_i : U_i \to M$ of M and every coordinate system $\tilde{\gamma}_j : V_j \to N$ on N the function $\tilde{\gamma}_j^{-1} \circ f \circ \gamma_i : U_i \to V_j$ is infinitely differentiable.*

Notice how the various pieces of the formal definition of a manifold, Definition 9, each play a role. The function $f \circ \gamma_i$ is a function of n real variables whose domain U_i is an open subset of $\mathbb{R}^n$ by condition 1, so we can use techniques of multivariable calculus to differentiate it. By condition 2 any point $p \in M$ is guaranteed to lie in some coordinate patch; i.e., there must be an i such that $p \in \gamma_i(U_i)$, so for any point p and any function f the definition has teeth. One might worry about points p that lie in more than one coordinate patch – what if different coordinate patches yielded inconsistent answers? – but condition 3 comes to the rescue. To see that differentiability does not depend on the choice of coordinate patch in the domain manifold M, note that if p lies in both $\gamma_i(U_i)$ and $\gamma_j(U_j)$, then by condition 3 the function $\gamma_i^{-1} \circ \gamma_j$ is differentiable. So, by the chain rule from multivariable calculus (see [MTW, Section 2.4]) and the equality $f \circ \gamma_j = (f \circ \gamma_i) \circ (\gamma_i^{-1} \circ \gamma_j)$ it follows that $f \circ \gamma_j$ is differentiable at $\gamma_j^{-1}(p)$ if and only if $f \circ \gamma_i$ is differentiable at $\gamma_i^{-1}(p)$. An almost identical argument shows that differentiability does not depend on the choice of coordinate patch in the target manifold N. So we get the same answer to the differentiability question no matter which coordinate system we use for calculation.

Before leaving this section behind the reader should note that we have defined "infinitely differentiable" without showing how to take derivatives. In other words, we have not said how to make sense of "the derivative of a function f at a point p." In contrast, in calculus we start by defining derivatives and then build the notion of "infinite differentiability." We have deliberately avoided defining derivatives of functions on manifolds because the concept requires some more sophisticated machinery than we have yet developed. To give some idea of the issues involved, let us try naively to differentiate the function $g : S^1 \to S^2$ considered earlier. The derivative of the expression for g in $x_1 y_1$-coordinates as a function of t_1 (see Equation 3.5) is

$$t_1 \mapsto \begin{pmatrix} -\sin t_1 \sin \frac{5\pi}{18} \\ \cos t_1 \sin \frac{5\pi}{18} \end{pmatrix}.$$

On the other hand, if we write g in $\phi_1 \theta_1$-coordinates as a function of t_1, we get

$$(0, 2\pi) \to (0, 2\pi) \times (0, \pi)$$

$$t_1 \mapsto \begin{pmatrix} t_1 \\ \frac{5\pi}{18} \end{pmatrix}.$$

The derivative of this function is

$$t_1 \mapsto \begin{pmatrix} 1 \\ 0 \end{pmatrix}.$$

If we try to evaluate these derivatives at a particular point on the circle, such as $\begin{pmatrix} 0 \\ 1 \end{pmatrix}$, which corresponds to $t_1 = \pi/2$, we get the answer $\begin{pmatrix} 0 \\ \sin \frac{5\pi}{18} \end{pmatrix}$ in one set of coordinates and $\begin{pmatrix} 1 \\ 0 \end{pmatrix}$ in the other. So although we could determine differentiability without fear that different coordinate choices would give different answers, we cannot similarly evaluate derivatives. Exercise 50 gives a clue to the resolution of the difficulty; for the actual definition of the generalization of the notion of derivative to functions between manifolds, see [dC, Proposition 2.7 and Definition 2.8] or [Wa, Section 1.22].

Exercise 50 *Calculate the Jacobian matrix of the change-of-coordinates function from $\phi_1 \theta_1$-coordinates to $x_1 y_1$-coordinates. Evaluate this Jacobian at a point on the 40th parallel corresponding to an arbitrary t_1 in the interval $(0, 2\pi)$. You will get a matrix depending on t_1. Apply the Jacobian matrix to the vector $\begin{pmatrix} 1 \\ 0 \end{pmatrix}$. What function of t_1 do you obtain?*

3.3   Vector Fields and Differential Forms on Manifolds

Vector fields and differential forms are crucial to the systematic exploitation of symmetry. In Chapter 2 we studied them on $\mathbb{R}^n$, but it is not hard to define them on any differentiable manifold. The only issue is compatibility across coordinate systems.

For instance, to define a vector field on a manifold M we can specify a coordinate vector field in each coordinate patch of M and require that two coordinate vector fields should "agree" on the overlap of the coordinate patches. More precisely, in the language of Definition 9 and Section 2.4, for each coordinate patch $\gamma_i : U_i \to M$ we have a vector field $\mathbf{v}_i$ defined on the open set $U_i \subset \mathbb{R}^n$, and we require that for any transition function Γ_{ji} we have $\mathbf{v}_j = (J_{\Gamma_{ji}})\mathbf{v}_i$. (In this notation subscripts denote different coordinate systems, not different components of one coordinate system. The formula holds for each fixed i and j.) To see why this technical condition captures the right intuitive notion, recall from Section 2.6 that the pushforward of a vector field is the proper way to change coordinates.

Exercise 51 *Show that the derivative of the function g from Section 3.2 with respect to the parameter t_1 on S^1 is a vector field by calculating its derivative $\mathbf{v}$ in each coordinate patch, calculating the Jacobian of each transition function Γ_{ji} and checking explicitly that $\mathbf{v}_j = (J_{\Gamma_{ji}})\mathbf{v}_i$.*

Similarly, one correct definition of a differential form on a manifold is a collection of coordinate differential forms $\{\alpha_1, \alpha_2, \dots\}$ defined on the open sets $\{U_1, U_2, \dots\}$ such that for any transition function Γ_{ji} we have $\alpha_i = \Gamma_{ji}^* \alpha_j$. For example, the gradient one-form of the height function on the sphere can be expressed in coordinates in each of the four coordinate systems introduced in Section 3.1. Explicitly, these coordinate expressions are $\alpha_1 = -\sin\theta_1 d\theta_1$, $\alpha_2 = -\sin\theta_2 d\theta_2$, $\alpha_3 = (1 - x_1^2 - y_1^2)^{-\frac{1}{2}}(x_1 dx_1 + y_1 dy_1)$ and $\alpha_4 = -(1 - x_2^2 - y_2^2)^{-\frac{1}{2}}(x_2 dx_2 + y_2 dy_2)$.

Exercise 52 *Calculate Γ_{21} and Γ_{31} for this example. Check that $\alpha_1 = \Gamma_{21}^* \alpha_2$ and $\alpha_1 = \Gamma_{31}^* \alpha_3$.*

From Exercise 52 one can easily argue (without further calculation) that the collection $\{\alpha_1, \alpha_2, \alpha_3, \alpha_4\}$ is indeed a differential form on the sphere. In every coordinate system on the sphere this one-form is the gradient one-form of the height function z, so its natural name is dz.

One can also think of a differential one-form on a manifold as a collection of real-valued linear functions on the tangent spaces to points of the manifold. Likewise, a differential two-form is a collection of antisymmetric bilinear forms, one on each tangent space. This more sophisticated point of view is particularly appealing because it is *coordinate-free*, i.e., it allows us to talk about differential forms without picking any particular coordinate system and without worrying about transition functions.

We will end this section with a formal definition of a symplectic manifold. In order to do so, we must define *closed* and *nondegenerate* two-forms.

In order to define nondegeneracy we must learn a bit more about differential two-forms. Recalling Exercise 24 and Section 2.3, we see that a differential two-form on an open set U in $\mathbb{R}^n$ can be thought of as an $n \times n$, antisymmetric matrix with entries that may vary.

Definition 12 *A two-form on an open set $U \in \mathbb{R}^n$ is called* nondegenerate *if the corresponding $n \times n$ matrix is nonsingular at every point of U.*

Exercise 53 *Show that every differential two-form on every open set $U \in \mathbb{R}^n$ can be written as a sum of terms of the form*

$$f\, dg \wedge dh,$$

where f, g and h are real-valued functions on U. Show that every differential three-form on every open set $U \in \mathbb{R}^n$ can be written as a sum of terms of the form

$$f\alpha \wedge \beta,$$

where f is a function, α is a one-form and β is a two-form.

Exercise 54 *Show that the differential two-form $\rho\, d\rho \wedge d\theta$ is nondegenerate for $(\rho, \theta) \in \mathbb{R}^+ \times \mathbb{R}$.*

Exercise 55 *Show that if M is an n-manifold and ω is a nondegenerate differential two-form on M, then n must be even.*

Defining closed forms takes more work. There are two ways to proceed, one algebraic and one geometric. Our goal here is not to do these ideas justice, but to develop them just enough to allow us to define a symplectic manifold rigorously. We start with the algebraic definition of closed forms. We first must define the *exterior derivative d*. In Section 2.2 we defined the gradient one-form df of a function f. So we can think of d as a (linear) transformation from the vector space of functions to the vector space of one-forms. We can extend the definition of d to include a linear transformation from one-forms to two-forms by defining $d(f\,dx) := df \wedge dx$ for any function f and any coordinate x. We can extend the definition of the exterior derivative d to two-forms by defining $d(\alpha \wedge \beta) := (d\alpha) \wedge \beta + \alpha \wedge (d\beta)$ for any one-forms α and

β and extending the definition linearly to all two-forms. Finally, we can extend the definition to two-forms on manifolds by checking that the forms all pull back properly under transition functions. For a careful, detailed exposition of these ideas, see Frankel [Fr, Chapter 2] or Warner [Wa, Chapter 2].

The result of applying d to a differential two-form is a differential three-form, that is, an antisymmetric trilinear form on the tangent space at each point. In coordinates, a differential three-form on an n-dimensional manifold is a function from the manifold to the set of antisymmetric trilinear forms on $\mathbb{R}^n$. Every three-form can be written as a sum of wedge products of one-forms and two-forms (see Exercise 53). In a more thorough study of differential geometry, differential three-forms and even differential n-forms (for many natural numbers n) play an important role. For our purposes, differential three-forms are important only as exterior derivatives of two-forms. We can now give the algebraic definition.

Definition 13 (algebraic) *A differential form ω is closed if and only if $d\omega = 0$.*

The geometric definition of closed two-forms involves integrating them on two-dimensional spheres. A two-dimensional sphere in $\mathbb{R}^n$ is the image of an infinitely differentiable function $f : S^2 \to \mathbb{R}^n$. To integrate a two-form ω over a sphere Σ we pull ω back to $\mathbb{R}^2$ via the composition $f \circ \tilde{\gamma}$, where $\tilde{\gamma}$ is the parametrization of the sphere defined in Equation 3.1. The pullback $(f \circ \tilde{\gamma})^*\omega$ is a two-form on $\mathbb{R}^2$ and hence can be written in coordinates as $g(\theta, \phi)d\theta \wedge d\phi$ for some function $g : \mathbb{R}^2 \to \mathbb{R}$. We then define

$$\int_\Sigma \omega := \int_0^{2\pi} \int_0^{\pi} g(\theta, \phi)d\theta d\phi.$$

The integral on the right-hand side is a garden variety double integral of the kind studied in multivariable calculus.

Definition 14 (geometric) *A differential two-form ω on an open set U in $\mathbb{R}^n$ is closed if and only if, for every two-dimensional sphere Σ lying inside an n-dimensional ball contained entirely in U we have*

$$\int_\Sigma \omega = 0.$$

In $\mathbb{R}^3$ we can interpret the geometric definition physically: in magnetostatics, the absence of magnetic monopoles implies that the magnetic flux integrates to zero over any closed surface. We can think of the magnetic flux as a two-form: it assigns to any two vectors at a point the magnetic flux through the parallelogram spanned by the two vectors, with a sign given by a right-hand rule. So a closed two-form is a flux without monopoles. We leave to the reader the enlightening task of supplying analogous geometric definitions of closed differential n-forms for other values of n.

Exercise 56 *State a geometric condition on a function f equivalent to the algebraic condition $df = 0$. State a geometric condition on a one-form α equivalent to the algebraic condition $d\alpha = 0$.*

For a more detailed geometric exposition of the exterior derivative, see Bamberg and Sternberg [BS, Volume I, Chapter 8 and Volume II, Chapter 15].

Once we have a definition for closed forms on $\mathbb{R}^n$ we can extend the definition to arbitrary manifolds. A two-form α on an arbitrary n-dimensional manifold M is *closed* if each of its coordinate forms is a closed form on $\mathbb{R}^n$.

Exercise 57 *Show that every two-form ω on a two-dimensional manifold M is closed.*

Exercise 58 *Show that every gradient one-form is closed. Give an algebraic argument and a geometric argument.*

We are at last ready to define a symplectic manifold rigorously.

Definition 15 *A symplectic manifold is a pair (M, ω), where M is a differentiable manifold and ω is a closed, nondegenerate differential two-form on M.*

It follows from Exercise 57 that any orientable two-dimensional surface with the area form is a symplectic manifold. Not all of these are phase spaces. For instance, the sphere and the torus (and, for that matter, all closed and bounded two-dimensional surfaces) fail to be phase spaces (see Figure 3.3). A closed

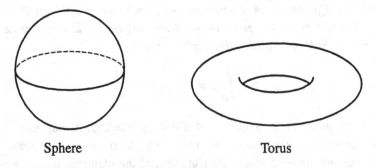

Sphere   Torus

Figure 3.3. Sphere and torus

and bounded surface in $\mathbb{R}^n$ is called *compact*. Phase spaces for classical, non-relativistic mechanical systems are never compact because the momenta can take arbitrarily large values. In the language of differential geometry, classical mechanical phase spaces are cotangent bundles (of the configuration spaces) and hence are not compact.

Readers already familiar with symplectic reduction should note that reductions of phase spaces, often called *reduced phase spaces*, can be compact. For instance, the reduced phase space of a completely integrable system is a point (see [AM, Examples 4.3.4]).

Exercise 59 *Check that all the phase spaces introduced in Chapter 2 satisfy Definition 15.*

4

Total Energy Functions
are Hamiltonian Functions

In introductory physics courses energy is usually introduced as a conserved quantity. In a closed system the energy is constant as the physical system moves in time. For any given physical system, each known conserved quantity provides an equation that can be very useful in the analysis of the system. For instance, conservation of energy for a ball of mass m thrown straight up into the air yields the equation $E = \frac{1}{2m} p^2 + mgr$, where E is the constant value of the energy, g is the constant strength of the downward gravitational force, p is the momentum of the ball and r is its height. From this equation alone, without using calculus, one can predict the maximum height of the ball from its initial position and momentum.

But energy is more than just a conserved quantity: it determines the equations of motion. All conceivable information about the system is hidden in the energy formula. Unpacking this information is a mathematical problem. In this chapter we will show how to derive the equations of motion from the energy function and the kinetics of phase space. In other words, we will show how a real-valued function on a symplectic manifold determines a vector field.

We will call our energy function a *Hamiltonian function* and denote it H. We call the corresponding vector field on phase space a *Hamiltonian vector field* and denote it X_H. The vector field corresponds naturally to a first-order system of differential equations on phase space, which in physical systems is equivalent to Newton's second law (force = mass × acceleration). The solution of this differential equation is the *Hamiltonian flow*. Physically the Hamiltonian flow represents possible physical motions.

In Section 4.1 we give a formula defining the Hamiltonian vector field in terms of the symplectic form and the Hamiltonian function. In Section 4.4 we study some special properties of Hamiltonian vector fields and flows. Examples fill the remaining sections.

4.1 Total Energy and Equations of Motion

In this section we give the recipe for determining the equations of motion from the total energy (i.e., the Hamiltonian function) and the kinetics of phase space (i.e., the symplectic form on the manifold). This recipe is known as *Hamilton's equation(s)*. We start with a well-known physical example and end with the general symplectic formulation.

Consider a particle moving on a line, with position r and momentum p. If the particle has mass m, then the kinetic energy of the particle is $\frac{1}{2m}p^2$. If the particle is moving under the influence of a conservative force (such as a spring force, discussed in more detail in Section 4.2) then there is a potential energy function $U : \mathbb{R} \to \mathbb{R}$ and the equation of motion is $\frac{d^2}{dt^2}r = -U'(r)$.

The total energy function, also known as the *Hamiltonian function*, for this example is

$$H := \frac{1}{2m}p^2 + U(r).$$

This total energy function H is important not only because it is *conserved* (i.e., $\frac{d}{dt}H = 0$ for any particular motion of the particle) but also, more fundamentally, because one can obtain the equations of motion from H via the recipe

$$\frac{dr}{dt} = \frac{\partial H}{\partial p} \quad \text{and} \quad \frac{dp}{dt} = -\frac{\partial H}{\partial r}. \tag{4.1}$$

These are known as *Hamilton's equations* for the energy function H.

Exercise 60 *Show that the conservation of H (i.e., the fact that $\frac{dH}{dt} = 0$) follows from Equations 4.1.*

Exercise 61 *Show that Equations 4.1 implies that $m\frac{d^2}{dt^2}r = -U'(r)$.*

Consider any function F defined on this phase space (such as temperature, or kinetic energy, or total energy). Then F is a function of r and p. We can calculate the time rate of change in F along a single particle's trajectory. Think (for a minute) of r and p as functions of time t (along a particular trajectory) and use the multivariable chain rule

$$\frac{d}{dt}F(r, p) = \frac{dr}{dt}\frac{\partial F}{\partial r} + \frac{dp}{dt}\frac{\partial F}{\partial p}.$$

So it makes sense to write

$$\frac{d}{dt} = \frac{dr}{dt}\frac{\partial}{\partial r} + \frac{dp}{dt}\frac{\partial}{\partial p}. \tag{4.2}$$

Notice that $\frac{d}{dt}$ is a vector field on the plane, as described in Section 2.3. We can rewrite the equations of motion (Equation 4.1) as

$$\frac{d}{dt} = \left(\frac{\partial H}{\partial p}\frac{\partial}{\partial r} - \frac{\partial H}{\partial r}\frac{\partial}{\partial p}\right).$$

The vector field $\frac{d}{dt}$ plays a special role and has a special name — the *Hamiltonian vector field*. It is often denoted X_H to emphasize its dependence on the Hamiltonian function H.

Let us write Hamilton's equations in the more formal language of symplectic geometry. The phase space is $\mathbb{R}^2$ with symplectic form $\omega = dp \wedge dr$. The gradient one-form of the Hamiltonian function can be written $dH = \frac{\partial H}{\partial p}dp + \frac{\partial H}{\partial r}dr$. We will plug X_H into the first slot of the symplectic form $dp \wedge dr$. Using Hamilton's equation (4.2) to justify the first equality we find that

$$
\begin{aligned}
dp \wedge dr (X_H, \cdot) &= dp \wedge dr \left(\frac{\partial H}{\partial p}\frac{\partial}{\partial r} - \frac{\partial H}{\partial r}\frac{\partial}{\partial p}, \cdot\right) \\
&= -\frac{\partial H}{\partial p}dp(\cdot) - \frac{\partial H}{\partial r}dr(\cdot) \\
&= -\frac{\partial H}{\partial p}dp - \frac{\partial H}{\partial r}dr = -dH.
\end{aligned}
$$

In this equation the dot is a placeholder for an arbitrary vector field. (We choose not to use the equivalent but more cumbersome notation: for any vector field v, we have $(dp \wedge dr)(X_H, v) = -dH(v)$.) If we write $\omega := dp \wedge dr$, then this last equation can be rewritten

$$\omega(X_H, \cdot) = -dH. \tag{4.3}$$

Exercise 62 *Show that Equation 4.3 implies Equations 4.1.*
This completes the proof that Equation 4.3 and Equations 4.1 are equivalent and justifies our using the name "Hamilton's equation(s)" for both.

To put Hamilton's equation in the form most familiar to modern symplectic geometers we must introduce one more concept. For the purposes of this book, the reader should feel free to think of this simply as a bit of fancy notation. We will use the symbol $\iota_{X_H}\omega$ to denote the differential one-form obtained by inserting the vector field X_H into the first slot of ω. This is sometimes called the *interior product of X_H and ω* (as in [Fr, Section 2.9]) or the *inner product of X_H and ω* (as in [AM, Definition 2.4]). In the matrix-based language of Sections 2.2 and 2.3, the formula $\iota_{X_H}\omega$ instructs us to multiply the row vector X_H^T by the matrix corresponding to the two-form ω to obtain a new row vector. Now we can rewrite Hamilton's equation as

$$\iota_{X_H}\omega = -dH. \tag{4.4}$$

Exercise 63 *Use the nondegeneracy of the symplectic form (see Definitions 12 and 15 in Section 3.3) to show that, given any infinitely differentiable real-valued function H on a phase space M with symplectic form ω, there is a unique vector field X_H satisfying Equation 4.4.*
Note: the solution of this exercise does not depend on techniques developed in Chapter 3.

According to this exercise, any infinitely differentiable function H on any symplectic manifold M with symplectic form ω generates one and only one Hamiltonian vector field. So it makes sense to define a *Hamiltonian system* to be a triple (M, ω, H), where M is a manifold, ω is a symplectic form on M, and H is an infinitely differentiable, real-valued function on M. If the symplectic form is $dp \wedge dr$, then Equations 4.1 are equivalent to Equation 4.4; for any other symplectic form one must use Equation 4.4.

Exercise 64 *Starting from Equation 4.4, derive Hamilton's equations in terms of the coordinates and the Hamiltonian function H for the phase space $\mathbb{R}^n$ with the canonical symplectic form introduced in Section 2.5.*

Note that the symplectic form plays a crucial role here: the same function H can induce different flows (X_H's) for different ω's. For instance, as we show in Section 4.3, the presence of a magnetic field (which affects the symplectic form but not the total energy of a charged particle) changes the motion of the particle without changing the Hamiltonian. It is the symplectic structure of phase space, in conjunction with the Hamiltonian function, that determines the flow of the system. In more physical language, it is the kinematics, in conjunction with the energy function, that determines the motion.

4.2 Particles on the Line

We start with two examples of physical systems whose phase spaces are the simplest nontrivial symplectic manifold, the two-plane $M = \mathbb{R}^2 = (r, p)$, with the area two-form $\omega = dp \wedge dr$. We studied this symplectic manifold in Section 2.1.

Consider first a particle of mass m moving on a line, subject to no forces. Such a particle is called a *free* particle. Our physical intuition (inherited from Newton) tells us that any free particle will travel with constant speed. The Hamiltonian function is

$$H = \frac{1}{2m}p^2.$$

Because the particle is free, we need only the kinetic energy term and no potential energy term in the Hamiltonian.

Let us practice using $\iota_{X_H}\omega = -dH$ to find the Hamiltonian vector field X_H. Keep in mind that X_H is just the time derivative $\frac{d}{dt}$ along trajectories. We write $X_H = x_r\frac{\partial}{\partial r} + x_p\frac{\partial}{\partial p}$ For the calculation, it helps to introduce an arbitrary vector field $v = v_r\frac{\partial}{\partial r} + v_p\frac{\partial}{\partial p}$. So x_r, x_p, v_r and v_p are real-valued functions of r and p that we have introduced in order to do an explicit calculation. We have

$$
\begin{aligned}
\iota_{X_H}\omega(v) &= dp \wedge dr(X_H, v) = dp \wedge dr\left(x_r\frac{\partial}{\partial r} + x_p\frac{\partial}{\partial p}, v_r\frac{\partial}{\partial r} + v_p\frac{\partial}{\partial p}\right) \\
&= x_p v_r - x_r v_p.
\end{aligned}
$$

On the other hand,

$$-dH(v) = -\frac{1}{m}p\, dp(v) = -\frac{1}{m}pv_p.$$

Since we must have $\iota_{X_H}\omega(v) = -dH(v)$ for every vector field v, we have, for every choice of the functions v_r and v_p, the equality $x_p v_r - x_r v_p = -\frac{1}{m}pv_p$. We can conclude that $x_p = 0$ and $x_r = \frac{1}{m}p$. Putting these together, we have

$$X_H = \frac{1}{m}p\frac{\partial}{\partial r}.$$

In less cumbersome notation, $\iota_{X_H}\omega = dp \wedge dr(x_r\frac{\partial}{\partial r} + x_p\frac{\partial}{\partial p}, \cdot) = -x_r dp + x_p dr$ and $-dH = -\frac{1}{m}p dp$. Hence, since dp and dr are linearly independent, we have $X_H = \frac{1}{m}p\frac{\partial}{\partial r}$.

Let us derive the equations of motion from the expression for X_H. Using Equation 4.2 and the fact that X_H is just another name for $\frac{d}{dt}$, we obtain

$$\frac{1}{m}p\frac{\partial}{\partial r} = \frac{dr}{dt}\frac{\partial}{\partial r} + \frac{dp}{dt}\frac{\partial}{\partial p}.$$

Since $\frac{\partial}{\partial r}$ and $\frac{\partial}{\partial p}$ are linearly independent, it follows that $\frac{d}{dt}r = \frac{1}{m}p$ and $\frac{d}{dt}p = 0$. In other words, a free particle travels with constant momentum along the line. This is equivalent to $\frac{d^2}{dt^2}r = 0$, the equation of motion for the free particle. So Equation 4.4 gives the correct equation of motion in this example.

Our second example of a physical system whose phase space is the plane is a particle moving on a line, subject only to the force of a spring (attached to the particle at one end and with equilibrium point at the origin). The reader should pause to imagine all the possible trajectories (behaviors) of such a particle. A common model for the spring force is $-kr$, where $k > 0$ is a constant. This is a very idealized model of a spring. Physically the spring constant k usually depends on r; the Hamiltonian formalism works for this case as well, but the calculations are more complicated. Of course, in actual practice, one can't ignore the dissipative effects of friction for which one needs more sophisticated mathematical machinery than that developed here. For simplicity of presentation, we stick to constant k. So the Newtonian equation of motion for our spring is $m\frac{d^2}{dt^2}r = -kr$.

The Hamiltonian function is

$$H = \frac{1}{2m}p^2 + \frac{k}{2}r^2.$$

The second term in the Hamiltonian is the potential energy due to the spring. Notice that the potential energy is larger the farther the particle is from the origin, which makes sense. If we write $X_H = x_r\frac{\partial}{\partial r} + x_p\frac{\partial}{\partial p}$, we find that $\iota_{X_H}\omega = -x_r dp + x_p dr$ and $dH = \frac{1}{m}pdp + krdr$. So Equation 4.4 implies that $X_H = \frac{1}{m}p\frac{\partial}{\partial r} - kr\frac{\partial}{\partial p}$. (See Figure 2.3 for the case $m = k = 1$.) Using Equation 4.2, we find that $\frac{d}{dt}r = \frac{1}{m}p$ and $\frac{d}{dt}p = -kr$. These equations together imply the Newtonian equation $m\frac{d^2}{dt^2}r = -kr$.

Exercise 65 *Apply the recipe given by Equation 4.4 to the Hamiltonian function $H := \frac{1}{2m}p^2 + mgr$ for a particle of mass m in a constant gravitational field of strength g. Show that the equation $\frac{d}{dt} = X_H$ implies that force = mass × acceleration, where the force has strength mg downward.*

4.3   Magnetism in Three-Space

We can use the Hamiltonian functions and symplectic forms to study the motion of a particle in three-space under the influences of conservative forces and a constant magnetic field. Because the magnetic field is not conservative we cannot hope to encode it in the total energy function. In this section we will see that we can encode it in the symplectic form ω so that Equation 4.4 yields the correct equations of motion for a charged particle.

First we will recall the equations of motion for such a particle. We know from elementary physics that the force created by the magnetic field is $\frac{q}{m}\mathbf{p}^T \times \mathbf{B}$, where $\frac{1}{m}\mathbf{p}^T$ is the velocity of the particle, q is the charge and $\mathbf{B} = (B_x, B_y, B_z)^T$ is the magnetic field vector. If all other forces acting on the particle are conservative, then they can be combined into one potential function $U : \mathbb{R}^3 \to \mathbb{R}$ such that the total of the other forces is minus the gradient vector of U, which we denote $-\nabla U$. So the equation of motion for the particle is

$$m\frac{d^2 r}{dt^2} = \frac{q}{m}\mathbf{p}^T \times \mathbf{B} - \nabla U.$$

Next let us check that Equation 4.4 can yield the same law of motion. We take for our symplectic form

$$\omega := dp_x \wedge dr_x + dp_y \wedge dr_y + dp_z \wedge dr_z$$
$$+q(B_x dr_y \wedge dr_z + B_y dr_z \wedge dr_x + B_z dr_x \wedge dr_y).$$

We choose the Hamiltonian function $H := \frac{1}{2m}\mathbf{p}^2 + U(\mathbf{r})$, where $\mathbf{r}$ is the column three-vector denoting the position of the particle. So

$$-dH = -\frac{1}{m}(p_x dp_x + p_y dp_y + p_z dp_z) - dU,$$

where dU is the gradient one-form of the function U, as defined in Section 2.3. Setting $\iota_{X_H}\omega$ equal to $-dH$ we obtain

$$X_H = \frac{1}{m}p_x\frac{\partial}{\partial r_x} + \frac{1}{m}p_y\frac{\partial}{\partial r_y} + \frac{1}{m}p_z\frac{\partial}{\partial r_z}$$
$$+ \left(\frac{q}{m}(p_y B_z - p_z B_y) - \frac{\partial U}{\partial r_x}\right)\frac{\partial}{\partial p_x}$$
$$+ \left(\frac{q}{m}(p_z B_x - p_x B_z) - \frac{\partial U}{\partial r_y}\right)\frac{\partial}{\partial p_y}$$
$$+ \left(\frac{q}{m}(p_x B_y - p_y B_x) - \frac{\partial U}{\partial r_z}\right)\frac{\partial}{\partial p_z}.$$

Exercise 66 *Derive this last formula for X_H.*

It follows that $\frac{d}{dt}\mathbf{r} = \frac{1}{m}\mathbf{p}^T$ and $m\frac{d^2}{dt^2}\mathbf{r} = \frac{d}{dt}\mathbf{p}^T = \frac{q}{m}(\mathbf{p}^T \times \mathbf{B}) - \nabla U$, as desired. So although magnetic forces are not conservative, we can discuss them in the framework of Hamiltonian mechanics if we think of the magnetic force as part of the kinematics of the particle.

Exercise 67 *Consider a charged particle constrained to move in a plane, subject to a magnetic field of strength $B(x, y)$ at each point (x, y) and subject to a conservative force with potential function $U(x, y)$. Write down the equations of motion for this particle and find a two-form ω and a function H which show that this motion is a Hamiltonian system.*

4.4   Energy Conservation and Applications

In this section we will show that the Hamiltonian flow preserves both the Hamiltonian function and the symplectic form. We will work out examples and show how to use the level sets of the Hamiltonian function to help predict behavior.

We know from introductory physics that the energy should be constant along trajectories. This fact generalizes to all Hamiltonian flows on symplectic manifolds:

Theorem 1 *Let (M, ω, H) be a Hamiltonian system. Then $X_H H = 0$.*

Proof. Suppose $m(t)$ is a solution of the equations of motion. Then we find that for any t the vector $m'(t)$ is equal to the vector field X_H evaluated at the point $m(t)$. So at any point $m(t)$ we have

$$X_H H = \frac{d}{dt}H(m(t)) = dH\left(m'(t)\right) = dH(X_H) = -\iota_{X_H}\omega(X_H)$$
$$= -\omega(X_H, X_H) = 0,$$

since ω is a two-form and hence is antisymmetric (see Section 2.3).   ◊

We can use this *conservation of the Hamiltonian by the Hamiltonian flow* to draw useful pictures, because it implies that the *orbits* of the system must lie inside level sets of H. (An *orbit* is a set of all points in phase space that the system passes through during one particular motion. In other words, it is the set of all points on one particular trajectory.) One of the beautiful features of Hamiltonian systems is that we can get information about orbits of the differential equations of motion (usually hard to solve) by solving the algebraic equation $H = $ constant (usually easier to solve).

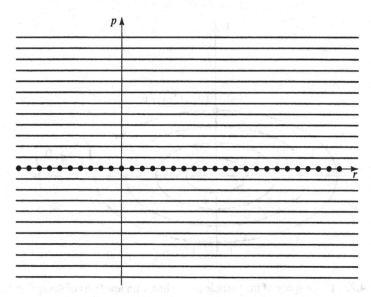

Figure 4.1. Phase space of the particle on the line with level sets of the force-free Hamiltonian

In the analysis of the motion of a free particle on the line (introduced in Section 4.2, conservation of the Hamiltonian by the Hamiltonian flow tells us that orbits must lie inside sets of the form

$$\frac{1}{2m}p^2 = \text{constant}.$$

Since the motion is continuous, it follows that each orbit is contained in a line $p = \text{constant}$. See Figure 4.1. Notice that not every orbit is an entire line. The r-axis ($p = 0$) is made up of single-point orbits representing motionless particles. All other orbits are entire lines representing particles moving at constant velocities.

If we wish to find bona fide solutions to our differential equation, i.e., if we wish to know not only the orbit of a trajectory but the trajectory itself (i.e., the position as a function of time, not just the set of positions), we have a little more work to do. We must solve one differential equation, which is usually simpler than solving the original system of differential equations. Although in this particular example it is just as easy (in fact, it's easier!) to solve the original system ($\frac{d}{dt}r = \frac{1}{m}p$ and $\frac{d}{dt}p = 0$), we will proceed pedantically, following the general method. Fix a value H_0 of the Hamiltonian and consider an orbit of energy H_0, i.e., an orbit contained in the set $\{(r, p) : \frac{1}{2m}p^2 = H_0\}$. We use the algebraic equation to reduce our system of two scalar differential equations to

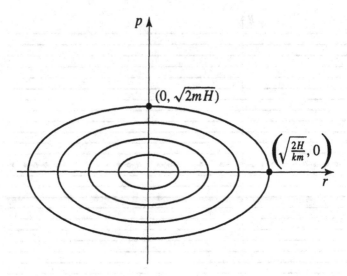

Figure 4.2. Phase space of the particle on the line with level sets of the spring Hamiltonian

one scalar differential equation:

$$\frac{dr}{dt} = \frac{1}{m}p = \pm\sqrt{\frac{2H_0}{m}}.$$

If the orbit lies on a line above the r-axis we take the plus sign; otherwise we take the minus sign. The equation above is easily integrated: the r-coordinate of a solution (a possible trajectory of a particle) is

$$r(t) = \pm\sqrt{\frac{2H_0}{m}}t + r(0) = \frac{1}{m}tp(0) + r(0).$$

From our algebraic equation $\frac{1}{2m}p^2 = H_0$ we see that p is constant, so $p(t) = p(0)$.

Next we apply conservation of the Hamiltonian to the example of the spring force on a particle on the line, introduced in Section 4.2.

Exercise 68 *Check explicitly that* $X_H H = 0$ *for the spring, i.e., that the Hamiltonian flow preserves the Hamiltonian.*

We draw the level sets of H in phase space; they are ellipses (see Figure 4.2). Note that if k is large the ellipses are tall and skinny, while if k is close to 0 then the ellipses are short and wide. If $k = \frac{1}{m}$ the ellipses degenerate to circles. Note that if $m = k = 1$ we get the vector field of Figure 2.3. Because the flow preserves the Hamiltonian, each solution of the system must lie entirely within one ellipse in phase space. In fact, each orbit of the system is an entire ellipse in phase space.

Exercise 69 *Consider a particle on a line moving according to the Hamiltonian $H = \frac{1}{2m}p^2 - \frac{k}{r}$, where $k > 0$ is a constant. This is the Hamiltonian function for the two-body problem in case the system has no angular momentum, i.e., in case the relative displacement of the bodies is parallel to their relative momentum. Use the equations of motion and the conservation of the Hamiltonian to show that on an orbit where $H = H_0$ we have*

$$\frac{d}{dt}r = \pm\sqrt{\frac{2}{m}(H_0 + \frac{k}{r})}.$$

Solve this equation by separation of variables and interpret the solutions physically.
The answer is important in Section 8.4.

Another fundamental theorem of Hamiltonian mechanics says that *the Hamiltonian flow preserves the symplectic form.* To state it formally we must first introduce a formal description of the Hamiltonian flow. We define the *Hamiltonian flow* to be a collection of functions $\Gamma_t : M \to M$, where t ranges over all real numbers, with the property that for each $m \in M$ the time-dependent function $\Gamma_t(m)$ solves the equations of motion and Γ_0 is the identity function on M. In equations, this means that for all $t \in \mathbb{R}$ and all $m \in M$ we have $X_H(\Gamma_t(m)) = \frac{d}{dt}(\Gamma_t(m))$ and $\Gamma_0(m) = m$.

Theorem 2 *Let (M, ω, H) be a Hamiltonian system. Then for each $t \in \mathbb{R}$ we have*

$$\Gamma_t^*\omega = \omega.$$

The proof of this theorem requires more machinery than we will develop in this book. One can make sense of the expression $X_H\omega = 0$ and calculate the preservation of the symplectic form directly, without explicitly solving for the Hamiltonian flow. The necessary technology is the Lie derivative, which the ambitious reader can study in Frenkel [Fr, Section 4.2] or in Abraham and Marsden [AM, Section 2.2]. For a proof of this theorem, see [AM, Prop. 3.3.4] or [Ar89, p. 204].

Let us see what this means in the context of the free particle. For any fixed time t the map is

$$\Gamma_t: \quad \mathbb{R}^2 \;\to\; r e^2$$
$$\begin{pmatrix} r \\ p \end{pmatrix} \mapsto \begin{pmatrix} r + \frac{1}{m} pt \\ p \end{pmatrix}.$$

We use the techniques of Section 2.6 to pull the symplectic form $\omega = dp \wedge dr$ back along the map Γ_t.

$$\Gamma_t^*\omega = (\Gamma_t^*dp) \wedge (\Gamma_t^*dr) = dp \wedge (dr + \frac{t}{m}dp) = dp \wedge dr = \omega.$$

In other words, the symplectic form ω is preserved under the flow.

Exercise 70 *Use Exercises 13 and 36 to give another, more geometric, proof of the preservation of ω under the flow.*

Exercise 71 *Show that the equation $X_H H = 0$ is equivalent to the equation $\Gamma_t^* H = H$.*

To clarify what preservation of the symplectic form under the flow means, we do a (non-Hamiltonian!) example of a flow not preserving the symplectic form. Consider the flow

$$\Gamma_t: \quad \mathbb{R}^2 \to \mathbb{R}^2$$
$$\begin{pmatrix} r \\ p \end{pmatrix} \mapsto \begin{pmatrix} re^t \\ pe^t \end{pmatrix}.$$

Then, for any given t, we find that

$$\Gamma_t^*\omega = (\Gamma_t^*dp) \wedge (\Gamma_t^*dr) = d(pe^t) \wedge d(re^t) = e^{2t} dp \wedge dr = e^{2t}\omega,$$

so the symplectic form is not preserved. See Figure 4.3. It follows that this flow is not the flow of any Hamiltonian system with the standard symplectic form on $\mathbb{R}^2$.

Exercise 72 *Find a symplectic form on $\mathbb{R}^2 \backslash (\{0\} \cup \mathbb{R}^+)$ that is preserved under this flow. Find a Hamiltonian function H for which this Γ_t is the Hamiltonian flow.*

We will check that the Hamiltonian flow of the a particle on the line with a spring force preserves the symplectic form $dr \wedge dp$ by using an explicit expression for Γ_t.

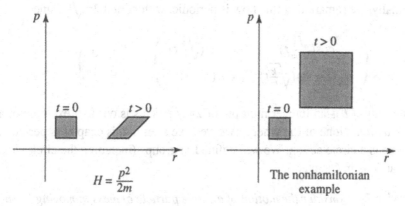

Figure 4.3. Symplectic form preserved or not preserved

Exercise 73 *Show that*

$$
\Gamma_t : \begin{pmatrix} r \\ p \end{pmatrix} \mapsto \begin{pmatrix} \cos(\sqrt{\frac{k}{m}}t) & \frac{1}{\sqrt{km}}\sin(\sqrt{\frac{k}{m}}t) \\ -\sqrt{km}\sin(\sqrt{\frac{k}{m}}t) & \cos(\sqrt{\frac{k}{m}}t) \end{pmatrix} \begin{pmatrix} r \\ p \end{pmatrix}
$$

is the Hamiltonian flow for a particle on the line subject to a spring force.

We remark that Hamiltonian flows cannot always be written in matrix form. In this case, however, the differential equations are linear, so a matrix solution exists.

One way to visualize this flow is to rewrite it:

$$
\begin{pmatrix} r \\ p \end{pmatrix} \mapsto \begin{pmatrix} \frac{1}{\sqrt{km}} & 0 \\ 0 & 1 \end{pmatrix} \begin{pmatrix} \cos(\sqrt{\frac{k}{m}}t) & \sin(\sqrt{\frac{k}{m}}t) \\ -\sin(\sqrt{\frac{k}{m}}t) & \cos(\sqrt{\frac{k}{m}}t) \end{pmatrix} \begin{pmatrix} \sqrt{km} & 0 \\ 0 & 1 \end{pmatrix} \begin{pmatrix} r \\ p \end{pmatrix}.
$$

Geometrically, the flow at time t in phase space is effected by first scaling the r-axis by a factor of $\sqrt{km}$ (the rightmost 2×2 matrix on the right-hand side), which takes the orbits to circles; second, rotating these circles clockwise through an angle $\sqrt{\frac{k}{m}}t$ (the middle matrix); and finally unscaling the r-axis back to its original scale (the leftmost matrix).

Let us check that the flow preserves the symplectic form. Notice that for each t the function Γ_t is a linear function from $\mathbb{R}^2$ to $\mathbb{R}^2$. Because the determinant of the matrix representing Γ_t is 1, we know that Γ_t is area-preserving. See Exercise 13. So the flow preserves the symplectic form.

Finally, we remark that this flow is periodic, with period $2\pi\sqrt{\frac{m}{k}}$, since

$$
\begin{pmatrix} \cos(\sqrt{\frac{k}{m}}t) & \frac{1}{\sqrt{km}}\sin(\sqrt{\frac{k}{m}}t) \\ -\sqrt{km}\sin(\sqrt{\frac{k}{m}}t) & \cos(\sqrt{\frac{k}{m}}t) \end{pmatrix} = \begin{pmatrix} 1 & 0 \\ 0 & 1 \end{pmatrix}
$$

if and only if t is an integer multiple of $2\pi\sqrt{\frac{m}{k}}$. This is our first example of a *circle action*. None of the other flows we have seen in this chapter is periodic, so none is a circle action. We will define Lie groups (including the circle) and their actions in Chapter 5.

Exercise 74 *Consider the motion of a single particle of mass m moving in one dimension in a constant gravitational field of strength g. The Hamiltonian for this system is*

$$
H = \frac{1}{2m}p^2 + gmr.
$$

See Exercise 65 for the equations of motion.

1. *Solve the differential equations of motion to find the most general equation of motion of the particle (that is, $r(t) = \dots$). Give physical interpretations of the new constants that appear in the equation.*

2. *Sketch r versus t for several different trajectories.*

3. *Express r in terms of p and sketch r versus p for several different trajectories. Put arrows on your sketch to indicate the direction of motion.*

4. *Find a 2 × 2 matrix J in (r, p) coordinates such that $X_H = J\nabla H$.*

5. *Sketch the gradient field ∇H in phase space. Sketch the level sets of H. Compare with 3. What is the relationship between the gradient arrows on your graph to the arrows indicating direction of motion in the graph from 3? Relate this to X_H and the geometric meaning of the matrix J.*

6. *Check that the Hamiltonian flow preserves the symplectic structure.*

Next we present two examples of physical systems whose phase spaces are the cylinder of radius R, with the natural area form. Each is a planar pendulum, i.e., a particle of mass m constrained to move on a circle of radius R. On the first no forces act; on the second a constant force acts. We studied the underlying symplectic manifold in Section 2.4. Mathematically, the phase space is

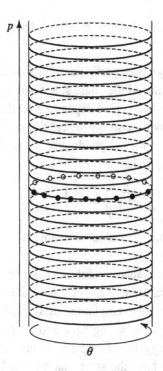

Figure 4.4.   Phase space of the free planar pendulum

an infinite cylinder $C_R := \{(p, R\cos\theta, R\sin\theta) : p \in \mathbb{R}, \theta \in \mathbb{R}\}$ of radius R with the area form $\omega = R\,dp \wedge d\theta$.

The first example on the cylinder is the free planar pendulum, with no forces acting (not even gravity). This system has a very simple set of physical behaviors: the particle can sit motionless at any point of the circle, or it can move at constant speed one way or the other around the circle. The Hamiltonian is pure kinetic energy:

$$H = \frac{1}{2m}p^2.$$

Each orbit is contained in a circle of constant p. For a picture of the orbits, see Figure 4.4. The distinct point orbits on the circle $p = 0$ correspond to motionless pendulums. Every other circle is a single orbit, representing a motion of the pendulum spinning at constant speed. To calculate this mathematically, note that the Hamiltonian vector field is $X_H = \frac{1}{mR}p\frac{\partial}{\partial\theta}$. The reader should check that the Hamiltonian flow is given by

$$\Gamma_t : \begin{pmatrix} \theta \\ p \end{pmatrix} \mapsto \begin{pmatrix} 1 & \frac{t}{mR} \\ 0 & 1 \end{pmatrix} \begin{pmatrix} \theta \\ p \end{pmatrix}.$$

Exercise 75 *Check that this flow preserves the Hamiltonian and the symplectic form.*

We remark that although each particular motion of the pendulum is periodic, the Hamiltonian flow on the whole phase space is not periodic. Each particular motion is periodic, but with a period depending on the motion. Notice that for the particle on the line with a spring force, the period was the same for all possible behaviors. Not so for the pendulum. Mathematically, we see that no choice of t (except $t = 0$) yields

$$\begin{pmatrix} 1 & \frac{t}{mR} \\ 0 & 1 \end{pmatrix} = \begin{pmatrix} 1 & 0 \\ 0 & 1 \end{pmatrix}.$$

(Looking ahead to Chapter 5, we say that this flow is an action of the line $(\mathbb{R}, +)$, not of the circle.)

Our next example is a planar pendulum under the influence of a uniform force of strength g parallel to the plane of the pendulum. The behavior is more complex than in the previous example: the mass can be motionless at the bottom or top of the circle, it can swing back and forth around the bottom, or it can, if it has enough energy, travel round and round the circle in either direction. Mathematically, we have the same phase space as before but a different Hamiltonian:

$$H = \frac{1}{2m}p^2 + Rmg(1 - \cos\theta).$$

Note that the bottom of the pendulum has the minimum potential energy and so corresponds to values of θ that are multiples of 2π. We can use the Hamiltonian to draw a picture of the orbits; see Figure 6.2. It is instructive to compare this picture with the standard picture of the phase plane drawn in most

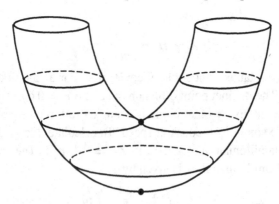

Figure 4.5. Phase space of the planar pendulum with gravity

undergraduate treatments of the pendulum (see, for instance, Simmons [Si, Section 63]). The reader may want to check that the Hamiltonian vector field is $X_H = \frac{1}{mR} p \frac{\partial}{\partial \theta} - mg \sin \theta \frac{\partial}{\partial p}$. The explicit integration of this flow (i.e., the closed-form solution calculated from the initial conditions) involves elliptic integrals. This problem is analyzed in detail in [Si, pp. 30 ff. and pp. 482 ff.] and in [Ar78, Example 12.9]. For a more sophisticated treatment, see the introduction to [CB]. This problem is one of the fundamental examples in *topological dynamics*, the study of how orbits fit together. See [AM, Section 4.5].

Exercise 76 *Let T be a positive real number. Find all Hamiltonian functions on the cylinder whose associated Hamiltonian flows rotate the cylinder steadily with period T.*

Exercise 77 *Let M be the surface of the sphere of radius R centered at the origin in $\mathbb{R}^3$. Let the symplectic form ω be the area form. Let the Hamiltonian function H be height. Describe the flow X_H geometrically.*
Hint: do this problem in cylindrical, yes cylindrical, coordinates.

The preservation of the symplectic form has applications in numerical analysis of Hamiltonian systems. Standard numerical techniques (such as Euler's method or Runge-Kutta algorithms) can easily make qualitatively wrong predictions for Hamiltonian systems. If these methods are altered to make sure that the numerically calculated flow preserves the symplectic form, they perform significantly better. For a basic example of this phenomenon and an in-depth discussion of the modified algorithms (called *symplectic integrators*), see Sanz-Serna and Calvo [SSC, Chapter 6].

4.5   An Example from the Two-Body Problem

Three Hamiltonian systems will arise in our mathematical analysis of the two-body problem in Chapter 8: the original Hamiltonian H_0 on the original twelve-dimensional phase space (M_0, ω_0), the Hamiltonian H_1 on the six-dimensional phase space (M_1, ω_1) obtained by exploiting conservation of linear momentum and the final Hamiltonian H_2 on the two-dimensional phase space (M_2, ω_2) obtained by exploiting conservation of angular momentum. In this section we will calculate the third Hamiltonian vector field X_{H_2} explicitly. For the derivation of this Hamiltonian and the precise physical interpretations of the coordinates and constants, see Chapter 8. Some readers may wish to skip this section until they have read enough of that chapter.

Our phase space is the half-plane $M_2 = \mathbb{R}^+ \times \mathbb{R} = \{(\rho, \sigma) : \rho > 0, \sigma \in \mathbb{R}\}$, with symplectic form $\omega_2 = d\sigma \wedge d\rho$. The coordinate ρ represents the distance between the two bodies and σ represents (roughly) the component of the momentum of one of the bodies parallel to the line joining the bodies. The Hamiltonian is

$$H_2 = \frac{|\lambda|^2}{2M} + \frac{1}{2\mu}\sigma^2 + \frac{|\tilde{L}|^2}{2\mu}\rho^{-2} - GM\mu\rho^{-1},$$

where everything but σ and ρ is constant. We can use the equation $\iota_{X_{H_2}}\omega_2 = -dH_2$ to find the vector field X_{H_2}. We write $X_{H_2} = x_\rho \frac{\partial}{\partial\rho} + x_\sigma \frac{\partial}{\partial\sigma}$ and find x_ρ and x_σ. It helps to introduce an arbitrary vector field $v = v_\rho \frac{\partial}{\partial\rho} + v_\sigma \frac{\partial}{\partial\sigma}$. We have

$$\iota_{X_{H_2}}\omega_2(v) = d\sigma \wedge d\rho(X_{H_2}, v) = d\sigma \wedge d\rho\left(x_\rho \frac{\partial}{\partial\rho} + x_\sigma \frac{\partial}{\partial\sigma}, v_\rho \frac{\partial}{\partial\rho} + v_\sigma \frac{\partial}{\partial\sigma}\right)$$

$$= x_\sigma v_\rho - x_\rho v_\sigma.$$

On the other hand,

$$dH_2 = \frac{\sigma}{\mu}d\sigma + \left(GM\mu\rho^{-2} - \frac{|\tilde{L}|^2}{\mu}\rho^{-3}\right)d\rho,$$

so we must choose x_ρ and x_σ to satisfy

$$x_\sigma v_\rho - x_\rho v_\sigma = -\left(\frac{\sigma}{\mu}v_\sigma + \left(GM\mu\rho^{-2} - \frac{|\tilde{L}|^2}{\mu}\rho^{-3}\right)v_\rho\right)$$

for any arbitrary v_ρ and v_π. It follows that the Hamiltonian vector field corresponding to the Hamiltonian system (M_2, ω_2, H_2) is

$$X_{H_2} = \frac{\sigma}{\mu}\frac{\partial}{\partial\rho} + \left(\frac{|\tilde{L}|^2}{\mu}\rho^{-3} - GM\mu\rho^{-2}\right)\frac{\partial}{\partial\sigma}.$$

5

Symmetries are Lie Group Actions

Every student of physics knows how important symmetries are. Calculations are easier and more enlightening in coordinates that recognize the symmetry of the underlying problem. For instance, calculating the volume of a cylinder is possible in spherical or Cartesian coordinates, but it is much easier in cylindrical coordinates.

Physically, a symmetry of a mechanical system is a change of coordinates that does not change the formula for the differential equations of motion. Another way to say this is to imagine two observers to whom the system looks identical. For instance, we could choose a coordinate system and sit at the origin, waiting for particles to whiz by. A friend of ours might sit at another point and introduce her own coordinate axes, whose origin coincides with her position. See Figure 5.1. Although our friend will experience any particular particle's motion differently (e.g., she might catch a baseball that goes nowhere near us) the set of all possible motions looks the same to her as it does to us (e.g., we are just as liable to find ourselves in the path of a baseball as she is). The change from her coordinates to our coordinates is a symmetry of the physics of baseballs.

In this chapter we will see that symmetries provide natural examples of mathematical *group actions*. We will define matrix Lie groups and abstract Lie groups, as well as group actions and orbits, and study several concrete examples. We will use orbits of group actions to construct useful quotient spaces.

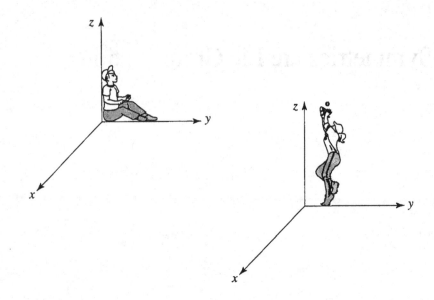

Figure 5.1.   Our coordinates and our friend's coordinates

Finally, we will apply all this mathematical technology to Hamiltonian systems. Along the way we will develop some efficient new notation.

5.1 From Symmetries to Groups

Physicists discuss the symmetries of a given physical system by imagining all possible observers who observe the same physical laws. For an introduction to this idea written for the general public, see [Z, Chapter 2] or [Gr, beginning of Chapter 7]. This is an application of what mathematicians call "group theory." (What physicists call "group theory" is a different but related subject that mathematicians refer to as "representation theory.") A *group* is defined to be a set with a multiplication operation that satisfies certain axioms.

Definition 16 *A group* $(G, \cdot)$ *is a set G with an operation* $G \times G \to G$ *denoted by* "$\cdot$" *and satisfying*

1. *Associativity: for all* g_1, g_2 *and* g_3 *in G we have* $(g_1 \cdot g_2) \cdot g_3 = g_1 \cdot (g_2 \cdot g_3)$.

2. *Existence of an Identity Element: there is an element I in G such that, for all* $g \in G$ *we have* $I \cdot g = g \cdot I = g$.

3. *Existence of Inverses: for each* $g \in G$ *there is an element, denoted* g^{-1}, *such that* $gg^{-1} = g^{-1}g = g$.

The requirement that the product of any two elements of G is itself an element of G is known as the *closure of G under multiplication*. For an introduction to groups at the level of an undergraduate abstract algebra course, see Artin [Art, Chapter 2].

There is a standard recipe for extracting a group from a set of equivalent observers. Imagine that you and a friend are two of the observers. You each have a copy of the mathematical phase space (M, ω) that describes the physical system. Now use physical observations to construct a function from M_0 to itself as follows. Given a point $q \in M$, there is a corresponding physical event from your point of view. For instance, if M is the phase space of a particle on the line (as in Section 2.1) then each point q in M corresponds to a particle at a particular position with a particular momentum. You create a particle with position and momentum given by q (from your perspective), and ask your friend to observe it. Her observation gives a unique point $g(q) \in M$ corresponding to that physical position and momentum (from her perspective). This defines the function $g : M \rightarrowtail M$.

If we consider the set of all possible equivalent observers (including ourself), we get a set G of functions from M to itself. We claim that this set, with the composition operation, forms a group. To prove the claim we must argue that composition of two elements of the set G yields another element of G and we must verify associativity, existence of the identity element and existence of inverses.

The hardest part is to show that G is closed under composition. Consider any two elements, g_1 and g_2, of the set G. These correspond to two friends, Friend 1 and Friend 2. Because Friend 2 is an equivalent observer, there must be a third friend, Friend 3, whose position relative to Friend 2 is exactly the same as Friend 1's position relative to us. If we create a particle with position and momentum q, then Friend 2 observes it as $g_2(q)$. Because of Friend 3's position relative to Friend 2, Friend 3 observes the particle to have position and momentum given by $g_1(g_2(q))$. So $g_3(q) = g_1(g_2(q))$ for every $q \in M$ or, in other words, $g_3 = g_1 \circ g_2$ and hence G is closed under composition.

Associativity always holds for composition of functions. The identity element is in G because in our model (and, we hope, in our reader's lives) we count as one of our own friends. We leave it to the reader to modify one of the above arguments to prove the existence of inverses.

Exercise 78 *Show that for any element $g \in G$ the inverse function g^{-1} also is an element of G.*

As an example, suppose that the system we wish to consider consists of vectors in $\mathbb{R}^2$. Suppose that sums, scalar multiples and lengths of these vectors are physically measurable. Then each observer determines an orthonormal coordinate system, and vice versa. The fact that equivalent observers must agree on sums and scalar multiples implies that each $g \in G$ must be a linear transformation of $\mathbb{R}^2$. If we pick one orthonormal basis for $\mathbb{R}^2$, then we can write each g as a matrix. Since equivalent observers agree on lengths, each matrix must be orthonormal. So the group of symmetries of this system is the set of 2×2 orthogonal matrices. This group is known as $O(2)$.

5.2   Matrix Lie Groups

Because linear vector spaces model so many physical phenomena, the most important symmetry groups are groups of linear transformations, known as *matrix Lie groups* or *linear Lie groups*. In this section we give the definition of matrix Lie groups and many examples. We encourage readers to supplement this section with Artin's chapter on matrix Lie groups [Art, Chapter 8].

Definition 17 *A* matrix Lie group *G* *is a nonempty set of invertible matrices with real or complex entries, together with the operation of matrix multiplication. The set G must be closed under the operations of matrix multiplication, matrix inversion and taking limits within the set of all invertible matrices.*

The last criterion of the definition means that if $\{g_n\}$ is a sequence of matrices, each of which lies in G, and if the sequence converges (entry by entry) to an invertible matrix g_∞, then g_∞ is an element of G. Note that because of the invertibility requirement G must contain only square matrices, and because of the multiplication requirement these matrices must all be the same size.

Fix any natural number n. Then the set of invertible $n \times n$ matrices with real entries is a matrix Lie group. It is called the *general linear group (over the reals)* and is denoted $GL(n, \mathbb{R})$. Let us check that it satisfies all the criteria of Definition 17. Since the identity $n \times n$ matrix is invertible, $GL(n, \mathbb{R})$ is nonempty. Because a matrix is invertible if and only if its determinant is nonzero, and because, for any matrices A and B we have $\det(AB) = \det(A) \cdot \det(B)$, it follows that $GL(n, \mathbb{R})$ is closed under matrix multiplication and matrix inversion. Finally, because all invertible $n \times n$ matrices lie in $GL(n, \mathbb{R})$ it is certainly closed under taking limits within the set of all invertible matrices. For example, the matrix Lie group $GL(1, \mathbb{R})$ is the same as $\mathbb{R} \setminus \{0\}$, with ordinary multiplication.

Not every natural set of matrices is a matrix Lie group. For example, fix any natural number n and consider the set of real matrices of trace 0. (Recall that the trace is the sum of the diagonal entries.) This is *not* a matrix Lie group because the product of two matrices of trace 0 need not have trace 0:

$$\begin{pmatrix} 1 & 0 \\ 0 & -1 \end{pmatrix} \begin{pmatrix} 1 & 0 \\ 0 & -1 \end{pmatrix} = \begin{pmatrix} 1 & 0 \\ 0 & 1 \end{pmatrix}.$$

It is even easier to note that not every matrix in the set is invertible. In fact, the set described is a *Lie algebra*. We will discuss Lie algebras in Chapter 6.

Exercise 79 *For any natural number n the set of invertible n × n matrices with rational entries is a group, denoted $GL(n, \mathbb{Q})$. Show (for every n) that $GL(n, \mathbb{Q})$ is a group but not a matrix Lie group.*

Exercise 80 *Show that every matrix Lie group is a group.*

Consider the set of positive real numbers $\mathbb{R}^+ := (0, \infty)$, with ordinary multiplication ($\times$). This is the matrix Lie group of 1×1 invertible matrices with positive real entries. It is not hard to check that $\mathbb{R}^+$ is nonempty and closed under matrix multiplication and matrix inversion. Let us check that it is closed under limits within the set of all invertible matrices. So suppose $\{a_n\}$ is a series of real, strictly positive numbers that converges to an invertible number a_∞. Then $a_\infty \neq 0$. Also, a_∞ cannot be negative, since a sequence of positive numbers cannot converge to a negative number. So a_∞ must be strictly positive, i.e., $a_\infty \in \mathbb{R}^+$. So $\mathbb{R}^+$ is closed under limits within the set of all invertible numbers.

We remark that there are many operations defined on $\mathbb{R}^+$. That is, we know how to do lots of things with positive real numbers: we can add them, multiply them, take the log of one base the other, etc. By writing $(\mathbb{R}^+, \times)$ we restrict our attention to multiplication alone. We forget all other operations. In general, the word group refers to a set on which only one algebraic operation is considered.

The set of real numbers, endowed with addition, denoted $(\mathbb{R}, +)$, is a group. The reader can easily check that it satisfies the criteria of Definition 16. Strictly speaking, $(\mathbb{R}, +)$ is not a matrix Lie group because the group operation is not matrix multiplication. However, $(\mathbb{R}, +)$ is *isomorphic* to a matrix Lie group: there is a matrix Lie group G and a differentiable function from $(\mathbb{R}, +)$ to G which is injective, surjective, has a differentiable inverse and respects the group operations. Consider the nonempty set

$$G := \{ \begin{pmatrix} 1 & t \\ 0 & 1 \end{pmatrix} : t \in \mathbb{R} \}.$$

Direct calculation of products and inverses shows that this set is closed under matrix multiplication and inversion. It is also easy to check that the set is closed under limits inside the set of all invertible matrices. The isomorphism between $(\mathbb{R}, +)$ and G given by

$$t \mapsto \begin{pmatrix} 1 & t \\ 0 & 1 \end{pmatrix} \tag{5.1}$$

is differentiable, since each entry of the matrix is a differentiable function of t. It is injective because no two different values of t correspond to the same matrix. It is surjective because each matrix in the group G corresponds to a value of t. To show that it respects the group operations we must check that 0 maps to the identity matrix and that the matrix corresponding to the sum of two real numbers is the product of the matrices corresponding to each summand. But it is easy to see that

$$0 \mapsto \begin{pmatrix} 1 & 0 \\ 0 & 1 \end{pmatrix}$$

and

$$t + u \mapsto \begin{pmatrix} 1 & t+u \\ 0 & 1 \end{pmatrix} = \begin{pmatrix} 1 & t \\ 0 & 1 \end{pmatrix} \begin{pmatrix} 1 & u \\ 0 & 1 \end{pmatrix}.$$

Checking differentiability of the inverse requires a modicum of differential geometry: how should we define differentiability of a function whose domain is a subset (and not even an open subset) of a space of matrices? Readers who do not want to stop and learn differential geometry may safely ignore issues of differentiability in this chapter. However, those readers who are familiar with manifolds and differentiable functions on manifolds (discussed in Chapter 3) should check differentiability of the inverse explicitly.

Exercise 81 (Optional) *Using the definitions in Chapter 3, show that G is a manifold and that the inverse of the function defined by the Formula 5.1 is a differentiable function.*

Exercise 82 *Show that $(\mathbb{R}, \times)$ is neither a group nor a matrix Lie group.*

Exercise 83 *Show that the Lie group $(\mathbb{R}, +)$ is isomorphic to the matrix Lie group $(\mathbb{R}^+, \times)$ via the correspondence*

$$(\mathbb{R}, +) \to \mathbb{R}^+$$
$$t \mapsto e^t.$$

An extremely important example of a matrix Lie group is the circle group, which arises in many different guises. One version of the circle group is the set of real orthogonal 2×2 matrices of determinant 1, also known as the *special orthogonal group (of 2×2 matrices)*, abbreviated as $SO(2)$. This set of matrices can be conveniently parametrized:

$$SO(2) = \left\{ \begin{pmatrix} \cos\theta & -\sin\theta \\ \sin\theta & \cos\theta \end{pmatrix} : \theta \in \mathbb{R} \right\}.$$

The group $SO(2)$ is isomorphic to the group S^1, defined to be the unit circle in the complex plane, with ordinary complex multiplication. This is a matrix Lie group of 1×1 matrices with complex entries.

To see that these two Lie groups are isomorphic, define the function

$$S^1 \to SO(2)$$

$$z \mapsto \begin{pmatrix} \Re z & -\Im z \\ \Im z & \Re z \end{pmatrix},$$

where $\Re z$ and $\Im z$ denote the real and imaginary parts of z. This function is injective, since two complex numbers are equal if and only if their real parts and imaginary parts agree. It is surjective because for any θ the point $\cos\theta + i\sin\theta$ lies in S^1. We leave it to the reader to practice complex and matrix multiplication by showing that

$$z_1 z_2 \mapsto \begin{pmatrix} \Re z_1 & -\Im z_1 \\ \Im z_1 & \Re z_1 \end{pmatrix} \begin{pmatrix} \Re z_2 & -\Im z_2 \\ \Im z_2 & \Re z_2 \end{pmatrix}.$$

As before, readers should feel free to either ignore or settle issues of differentiability.

Exercise 84 (Optional) *Using the definitions in Chapter 3, show that both S^1 and $SO(2)$ are manifolds and that the function above and its inverse are both infinitely differentiable.*

The next example plays an important role in the two-body problem, as we will see in Section 5.6. Consider Euclidean three-space, with addition as the group operation, $(\mathbb{R}^3, +)$. We write elements of this group as $\mathbf{g} = (g_x, g_y, g_z)^T$. This group is isomorphic to a matrix Lie group, as the reader can check from the explicit formula

$$\begin{pmatrix} g_x \\ g_y \\ g_z \end{pmatrix} \mapsto \begin{pmatrix} 1 & 0 & 0 & g_x \\ 0 & 1 & 0 & g_y \\ 0 & 0 & 1 & g_z \\ 0 & 0 & 0 & 1 \end{pmatrix}. \tag{5.2}$$

Exercise 85 *Show that the group* $(\mathbb{R}^+, 3)$ *is isomorphic to the group of invertible* 3×3 *matrices whose off-diagonal terms are all zero and whose diagonal terms are all positive.*

The rotation group of three-space also plays an important role in the analysis of the two-body problem. Consider the matrix Lie group

$$SO(3) := \{g \in GL(3, \mathbb{R}) : g^T g = I \text{ and } \det g = 1\}.$$

This group is called the *special orthogonal group (of* 3×3 *matrices)* because its elements are both special (determinant 1) and orthogonal ($g^T g = I$). In other words, it is the group of (orientation-preserving) rotations of three-space.

Physicists parametrize this group by what they call *Euler angles*. The equivalent mathematical statement is that element g of $SO(3)$ can be written as a product $g = g_1 g_2 g_3$, where g_1, g_2 and g_3 are each rotations around a coordinate axis. See [Go, Section 4.4] for an excellent discussion. Beware! Euler angles do not give an isomorphism between $SO(3)$ and three copies of the circle. There are some redundancies, i.e., there are some rotations that can be described in two essentially different ways by Euler angles.

Exercise 86 *Prove that the Euler angle function from* $S^1 \times S^1 \times S^1$ *to* $SO(3)$ *is not an isomorphism of Lie groups. Extra credit:* *prove that there is no Lie group isomorphism from* $S^1 \times S^1 \times S^1$ *to* $SO(3)$.

Exercise 87 *Show that for any* $g \in SO(3)$ *(the group of rotations) and any vectors* $\mathbf{r}, \tilde{\mathbf{L}} \in \mathbb{R}^3$, *we have*

$$g(\mathbf{r} \times \tilde{\mathbf{L}}) = (g\mathbf{r}) \times (g\tilde{\mathbf{L}}).$$

You can prove this by interpreting the cross-product geometrically. Another approach is:

1. *Prove the statement for any g which is a rotation around a coordinate axis.*

2. *Prove the statement for any rotation g, using Euler angles and part 1.*

5.3   Abstract Lie Groups (Optional)

For readers comfortable with differential geometry, here is an abstract definition of a Lie group.

Definition 18 *A* Lie group *is a manifold G endowed with an operation G* ×
G → *G (called* multiplication*) satisfying the axioms for group multiplication,*
such that both multiplication and inversion are infinitely differentiable func-
tions.

For example, let us check that for any natural number n, the set $GL(n, \mathbb{R})$ of
$n \times n$ matrices with nonzero determinant is a Lie group. We apply Definition 9
from Section 3.1 with one coordinate patch. There is an obvious linear map
from $\mathbb{R}^{n^2}$ to the set of $n \times n$ matrices — just arrange the entries of the n^2-
vector into n columns of length n. The requirement that the determinant be
nonzero pulls back to a strict polynomial inequality on $\mathbb{R}^{n^2}$, which, like any
strict polynomial inequality, is satisfied by an open subset U of $\mathbb{R}^{n^2}$. The image
of U under the map is all of $GL(n, \mathbb{R})$. So $GL(n, \mathbb{R})$ is a manifold. Since
each entry of a product of two matrices is a polynomial in the entries of the
factors, group multiplication is infinitely differentiable. Since by Cramer's rule
each entry of the inverse of a matrix is rational in the entries of the original
matrix, with denominator nonzero (since the determinant is nonzero), inversion
is infinitely differentiable. So $GL(n, \mathbb{R})$, with matrix multiplication, is a Lie
group.

Some might wonder why there is interest in an abstract theory of Lie groups
when most of the obviously useful examples are matrix Lie groups. (The fact
that matrix Lie groups are actually Lie groups follows from the fact that any
closed subgroup of a Lie group is itself a Lie group. For the proof, see Bröcker
and tom Dieck [BtD, Theorem 3.11].) There are at least two reasons to study
Lie groups abstractly. One is to appreciate the power of the interplay between
analysis and algebra, two subjects often separated in the curriculum. Another
is the possibility of generalizing intuitions gained from concrete matrix calculations to infinite-dimensional systems. In particular, this allows one to exploit
the symmetries of partial differential equations in a systematic way. For an introduction to results in mathematical physics, see Abraham and Marsden [AM,
Section 5.5] or Marsden and Ratiu [MR, Section 1.5, Section 1.6 and Chapter 3].

5.4 Group Actions

We now define the action of a group on a set and work out several examples.
Group actions are symmetries; hence we use the letter S.

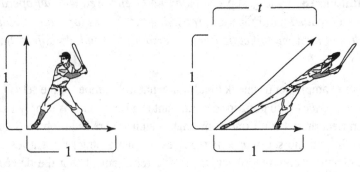

Figure 5.2. The action of $\begin{pmatrix} 1 & t \\ 0 & 1 \end{pmatrix}$ on $\mathbb{R}^2$.

Definition 19 *Let G be a group and let M be a set. An action of G on M is an assignment of a function $S_g : M \to M$ to each element $g \in G$ in such a way that*

1. *If I is the identity element of the group G, then S_I is the identity map, i.e., for any $m \in M$ we have $S_I(m) = m$.*

2. *For any $g, h \in G$, we have $S_g \circ S_h = S_{gh}$, i.e., for every $m \in M$, we have $S_g(S_h(m)) = S_{gh}(m)$.*

Strictly speaking, Lie group actions should satisfy certain differentiability properties in addition to the algebraic properties required by our definition; for a rigorous treatment, see [AM, Section 4.1] or [Br, Lecture 3].

We have already seen examples of an actions of $(\mathbb{R}, +)$ on manifolds. Recall the motion of a free particle along a line. The Hamiltonian flow is given explicitly by Formula 4.5. We have

$$\Gamma_t : \begin{pmatrix} r \\ p \end{pmatrix} = \begin{pmatrix} r + \frac{t}{m}p \\ p \end{pmatrix} = \begin{pmatrix} 1 & \frac{t}{m} \\ 0 & 1 \end{pmatrix} \begin{pmatrix} r \\ p \end{pmatrix}.$$

See Figure 5.2 for a picture of this action.

Let us verify the definition of an action in this example. In the language of the definition, we have $G = (\mathbb{R}, +)$ and $M = \mathbb{R}^2$. To check the first criterion, note that the identity element of the group is 0, and Γ_0 is indeed the identity matrix. To check the second criterion, note that for any t and u in $\mathbb{R}$ we indeed have

$$\Gamma_t \circ \Gamma_u = \begin{pmatrix} 1 & \frac{t}{m} \\ 0 & 1 \end{pmatrix} \begin{pmatrix} 1 & \frac{u}{m} \\ 0 & 1 \end{pmatrix} = \begin{pmatrix} 1 & \frac{t+u}{m} \\ 0 & 1 \end{pmatrix} = \Gamma_{t+u}.$$

This action is called *effective* because S_t is the identity map only for $t = 0$.

In fact, any Hamiltonian flow Γ_t is an action of $(\mathbb{R}, +)$, where the real parameter is interpreted as time. So even the spring Hamiltonian (analyzed in Section 4.2) gives us an action of the group $(\mathbb{R}, +)$. However, this action is not effective since at any time t which is an integer multiple of $2\pi\sqrt{\frac{m}{k}}$ we have

$$\begin{pmatrix} \cos(\sqrt{\frac{k}{m}}t) & \frac{1}{\sqrt{km}}\sin(\sqrt{\frac{k}{m}}t) \\ -\sqrt{km}\sin(\sqrt{\frac{k}{m}}t) & \cos(\sqrt{\frac{k}{m}}t) \end{pmatrix} = I.$$

Exercise 88 *Taking another point of view, we can think of this as an effective action of the circle group. Show that the Lie group of matrices of the above form is isomorphic to the circle group S^1.*

Because of this isomorphism we say that *the Hamiltonian flow of the spring is an action of the Lie group S^1* or *the Hamiltonian $\frac{1}{2}(\frac{p^2}{m} + kmr^2)$ generates a circle action on the plane.*

Every matrix Lie group, viewed as a group of linear transformations of a vector space, is an example of a Lie group action. Suppose G is a matrix Lie group of $n \times n$ matrices. For each $g \in G$ and each $\mathbf{v} \in \mathbb{R}^n$, define $S_g(\mathbf{v}) := g\mathbf{v}$. Then $S_I(\mathbf{v}) = I\mathbf{v} = \mathbf{v}$ for any $\mathbf{v} \in \mathbb{R}^n$. For each $g, h \in G$ and any $\mathbf{v} \in \mathbb{R}^n$ we have $S_g(S_h(\mathbf{v})) = S_g(h\mathbf{v}) = gh\mathbf{v} = S_{gh}(\mathbf{v})$. So this action satisfies Definition 19. This action is sometimes called the *natural action of a matrix Lie group.*

The matrix Lie group G acts naturally on the dual space $(\mathbb{R}^n)^*$ of row vectors as well.

Exercise 89 *Define $S_g(\mathbf{v}) := \mathbf{v}g^T$ for any $g \in G$ and any row vector $\mathbf{v}$ in $\mathbb{R}^n$. Show that S_g defines an action of G on $(\mathbb{R}^n)^*$.*

A combination of the natural action of $SO(3)$ on $\mathbb{R}^3$ and the action of $SO(3)$ on $(\mathbb{R}^3)^*$ plays an important role in the symplectic geometric analysis of the two-body problem in Chapter 8.

Matrix Lie groups also often act naturally on certain subsets of vector spaces. For example, because each element g of $SO(3)$ preserves lengths, it follows that for any column three-vector $\mathbf{v}$ in the unit sphere, the vector $g\mathbf{v}$ also lies in the unit sphere. So if we restrict the natural action $SO(3)$ to the unit sphere, we get an action of $SO(3)$ on the unit sphere.

The group $(\mathbb{R}^3, +)$ acts on a three-dimensional plane by translation as follows. For each $g \in (\mathbb{R}^3, +)$, define

$$S_{\mathbf{g}}(\mathbf{v}) := \mathbf{v} + \mathbf{g},$$

for each $\mathbf{v} \in \mathbb{R}^3$. It is easy to check that this is a group action. We can use the isomorphism of Formula 5.2 to exhibit this same action in a different way. Let M be the hyperplane $\{(x, y, z, 1)^T : x, y, z \in \mathbb{R}\}$ in $\mathbb{R}^4$. We have

$$
\begin{pmatrix} 1 & 0 & 0 & g_x \\ 0 & 1 & 0 & g_y \\ 0 & 0 & 1 & g_z \\ 0 & 0 & 0 & 1 \end{pmatrix} \begin{pmatrix} v_x \\ v_y \\ v_z \\ 1 \end{pmatrix} = \begin{pmatrix} v_x + g_x \\ v_y + g_y \\ v_z + g_z \\ 1 \end{pmatrix}.
$$

This action, called the *translation action of* $(\mathbb{R}^3, +)$ *on* $\mathbb{R}^3$, is related to linear momentum and plays an important role in the symplectic geometric analysis of the two-body problem in Chapter 8.

Exercise 90 *Which of the following define group actions on* $\mathbb{C}$? *Which actions are effective?*

1. $(\mathbb{R}, +)$ *acting by* $S_t : z \mapsto e^{it} z$.

2. $(\mathbb{R}, +)$ *acting by* $S_t : z \mapsto e^{i(t+1)} z$.

3. $GL(1, \mathbb{R})$ *acting by* $S_t : z \mapsto e^{it} z$.

4. $(\mathbb{R}, +)$ *acting by* $S_t : z \mapsto tz$.

5. $GL(1, \mathbb{R})$ *acting by* $S_t : z \mapsto tz$.

6. $(\mathbb{R}, +)$ *acting by* $S_t : z \mapsto e^{(1+i)t} z$.

Exercise 91 *Show that each of the following is a group action according to Definition 19.*

1. *The group* $\mathbb{Z}$ *of integers acting on* $\mathbb{R}$ *by addition:* $S_n(r) := r + n$.

2. *The matrix Lie group* $GL(n, \mathbb{R})$ *acting on itself by conjugation:* $S_g(h) := ghg^{-1}$.

3. *The matrix Lie group* $SO(3)$ *acting on itself by conjugation:* $S_g(h) := ghg^{-1}$.

4. *Any group acting on itself by left multiplication:* $S_g(h) := gh$.

5. *The action of* $SL(2, \mathbb{C})$ *(the group of* 2×2 *invertible matrices with complex entries) on the* Riemann *sphere* $\mathbb{C} \cup \{\infty\}$ *via linear fractional transformations (also known as* Möbius *transformations):*

$$
S_{\begin{pmatrix} a & b \\ c & d \end{pmatrix}}(z) := \frac{az + b}{cz + d}.
$$

5.5   Orbits of Group Actions

Symmetries often provide a mechanism for reducing the number of variables considered. For instance, if we are studying the electric field near the edge of two parallel conducting plates (see, for example, Feynman [Fe, Volume II, Section 6.10]) we can draw a two-dimensional picture that is an extremely good model of the three-dimensional system. This reduction is possible because translation parallel to the edge of the plates does not change the physics in any way. In other words, two points near the edge of the plates that differ only by a translation parallel to the edge are in some sense "the same point" and can be fairly and usefully represented by one and the same point on a two-dimensional picture. Mathematicians codify this sameness with the notions of *orbits of a group action* and *quotient spaces*.

Definition 20 *Let G be a group, let M be a set and let S_g (for each $g \in G$) be an action of G on M. For each $m \in M$ the orbit of the group through m is the set*

$$\mathcal{O}_m = \left\{ S_g(m) : g \in G \right\}.$$

Let us find all the orbits of the natural action of $SO(3)$ on $\mathbb{R}^3$. The orbit $\mathcal{O}_0$ through the zero vector $\mathbf{0}$ is $\{\mathbf{0}\}$ because the action is linear. Since each element of $SO(3)$ preserves lengths, the orbit $\mathcal{O}_\mathbf{v}$ through any nonzero three-vector $\mathbf{v}$ must be contained in the sphere of radius $|\mathbf{v}|$ around $\mathbf{0}$. On the other hand, it is not hard to see geometrically that, given any vector $\mathbf{w}$ with $|\mathbf{v}| = |\mathbf{w}|$, there is a rotation matrix carrying $\mathbf{v}$ to $\mathbf{w}$. Exercise 92 asks the reader to give an algebraic proof. So the orbits are the concentric spheres centered at the origin.

Exercise 92 *Given any vectors $\mathbf{v}$, $\mathbf{w}$ in $\mathbb{R}^3$ with $|\mathbf{v}| = |\mathbf{w}|$, give a formula for a matrix g in $SO(3)$ such that $g\mathbf{v} = \mathbf{w}$.*
Hint: Consider an orthonormal basis of $\mathbb{R}^3$ containing $\mathbf{v}/|\mathbf{v}|$ and another orthonormal basis containing $\mathbf{w}/|\mathbf{w}|$.

Exercise 93 *For each group action in Exercises 90 and 91, find all the orbits.*

The big payoff comes when we find ways to think of the set of orbits as a mathematical object in its own right. For instance, one can label each orbit of $SO(3)$ in $\mathbb{R}^3$ by a nonnegative number, the radius of the sphere. So the set of orbits in this case is a closed half line. This space is called the *quotient space of the action*.

Definition 21 *Let G be a group, let M be a set and let S_g (for each $g \in G$) be an action of G on M. The* quotient space *is*

$$M/G := \{\mathcal{O}_m : m \in M\}.$$

The symbol M/G is pronounced "M modulo G" or "M mod G." (We hope that any reader sophisticated enough to complain that we are using a group typographically on the right to denote the quotient of a left action will be broad-minded enough to appreciate the appeal of typography that matches verbal patterns.) There is a natural *quotient map*

$$q : M \to M/G$$
$$m \mapsto \mathcal{O}_m.$$

Quotient spaces often inherit some mathematical structure; at the same time, they can be more pathological than their parent spaces. For instance, while $\mathbb{R}^3$ is a manifold, its quotient space under the $SO(3)$ action, the closed half line, is not a manifold, as there are no nice coordinates at the point 0. Still, the topological structure of the closed half line is compatible with the topological structure of $\mathbb{R}^3$ in the sense that the quotient map is continuous. One of the major components of symplectic reduction is fact that certain quotient spaces are guaranteed to be not only manifolds but symplectic manifolds. This is the point of Theorem 3 in Chapter 8. In physical language, this theorem guarantees that if one reduces the number of variables correctly there will be good canonical coordinates on the lower dimensional space.

Exercise 94 *Consider the action of the group $(\mathbb{Z}, +)$ of integers acting on the real line $\mathbb{R}$ by translation: for each $g \in \mathbb{Z}$ and each $r \in \mathbb{R}$ we define $S_g(r) := r + g$. Show that $\mathbb{R}/\mathbb{Z}$ is isomorphic as a group to the circle group S^1. Optional extra credit: Show that $\mathbb{R}/\mathbb{Z}$ is isomorphic as a Lie group to the circle group S^1.*

5.6   Symmetries of Hamiltonian Systems

Symmetries are ubiquitous in mathematics and physics, so the concepts of this chapter have wide applicability. In this book we are concerned with one particular kind of problem, the analysis and solution of Hamiltonian systems. This section introduces the formal definition of a symmetry of a Hamiltonian system. We encountered this idea twice in Chapter 1 in disguised form. The twelve-dimensional system (Equations 1.1 and 1.2) of ordinary differential equations had translational symmetry, which allowed us to exploit center-of-mass coordinates, and the six-dimensional system (Equation 1.4) of ordinary differential equations had rotational symmetry, which allowed us to pass to polar coordinates (ρ, θ). Understanding the formal definition of a symmetry of a Hamiltonian system is an important step in lifting the disguise.

Definition 22 *A symmetry of a Hamiltonian system (M, ω, H) is a function $S : M \to M$ that preserves both the symplectic form ω and the Hamiltonian function H.*

The time flows discussed on the previous pages provide some examples of this phenomenon, as we saw in Theorems 1 and 2 of Section 4.4. However, a Hamiltonian system may have other symmetries in addition to the time flow.

We start with the translational symmetry of a free particle in three-space. We have

$$M = \mathbb{R}^6 = \{(r_x, r_y, r_z, p_x, p_y, p_z)\}$$
$$\omega = dp_x \wedge dr_x + dp_y \wedge dr_y + dp_z \wedge dr_z$$
$$H = \tfrac{1}{2m}(p_x^2 + p_y^2 + p_z^2).$$

Consider the translation action of $(\mathbb{R}^3, +)$ on M: for each $\mathbf{g} = (g_x, g_y, g_z)^T$ in $\mathbb{R}^3$, we define the function $S_{\mathbf{g}} : M \to M$ by

$$S_{\mathbf{g}}(\mathbf{r}, \mathbf{p}) = (\mathbf{r} + \mathbf{g}, \mathbf{p}).$$

This corresponds physically to an equivalent observer whose orientation is the same as the original observer's but whose coordinate origin lies at the point $-\mathbf{g}$ in the original observer's coordinate system. It is intuitively clear that in the absence of forces these two observers will see the same physical phenomena.

We will show mathematically that for any $\mathbf{g} \in \mathbb{R}^3$ the function $S_{\mathbf{g}}$ is a symmetry of the free particle phase space. First we show that $S_{\mathbf{g}}$ preserves the manifold M and its symplectic form ω. It is easy to see that $S_{\mathbf{g}}$ gives a one-to-one correspondence from $\mathbb{R}^6$ to itself. To show that $S_{\mathbf{g}}^* \omega = \omega$ we use the rules given in Section 2.6 and calculate

$$S_{\mathbf{g}}^* \omega = d(p_x) \wedge d(r_x + g_x) + d(p_y) \wedge d(r_y + g_y) + d(p_z) \wedge d(r_z + g_z)$$
$$= dp_x \wedge dr_x + dp_y \wedge dr_y + dp_z \wedge dr_z = \omega,$$

where the penultimate equality holds because $\mathbf{g}$ is constant and hence $dg_x = dg_y = dg_z = 0$. Next we check that $S_{\mathbf{g}}$ preserves the Hamiltonian, i.e., that

$S_g^* H = H$. But S_g changes only the values of $\mathbf{r}$, while H depends only on $\mathbf{p}$. So S_g preserves the Hamiltonian. We have shown that S_g is a symmetry.

Exercise 95 *Recall from Section 2.6 that $S_g^* H = H \circ S_g$ by definition. Check by explicit calculation that $S_g^* H = H$.*

Let us redo the calculation $S_g^* \omega = \omega$ in more compact notation. We think of $d\mathbf{p}$ as a row vector, and $d\mathbf{r}$ as a column vector and we write

$$d\mathbf{p} \wedge d\mathbf{r} = dp_x \wedge dr_x + dp_y \wedge dr_y + dp_z \wedge dr_z.$$

We are combining the wedge with matrix multiplication. We think of "d" as acting on each entry of the vectors: $d\mathbf{p} = d(p_x, p_y, p_z) = (dp_x, dp_y, dp_z)$ and

$$d\mathbf{r} = d \begin{pmatrix} r_x \\ r_y \\ r_z \end{pmatrix} = \begin{pmatrix} dr_x \\ dr_y \\ dr_z \end{pmatrix}.$$

If we write $\mathbf{g}$ as a column vector, we then have

$$S_g^* \omega = d(\mathbf{p}) \wedge d(\mathbf{r} + \mathbf{g}) = d\mathbf{p} \wedge d\mathbf{r} = \omega.$$

We will use this convenient notation whenever possible.

We summarize by saying that we have an action of the Lie group $(\mathbb{R}^3, +)$ on $(\mathbb{R}^6, d\mathbf{p} \wedge d\mathbf{r}, \frac{1}{2m}\mathbf{p}^2)$ preserving both the symplectic form and the Hamiltonian. Thus translation of the coordinate axes is a symmetry of the mechanical system. We will see in Chapter 7 that this symmetry is related to conservation of linear momentum.

Next, we consider the rotational symmetries of one free particle in three-space. (If a friend sits upside down or sideways to observe the particle, she'll see the same range of behaviors. Remember, a free particle experiences no forces, not even gravity.) Our Hamiltonian system is

$$M = \mathbb{R}^3 \times (\mathbb{R}^3)^* \cong \mathbb{R}^6$$
$$\omega = d\mathbf{p} \wedge d\mathbf{r}$$
$$H = \frac{1}{2m}|\mathbf{p}|^2.$$

We define an action of $SO(3)$ on M as follows. For each $g \in SO(3)$, define

$$S_g : M \to M, \; (\mathbf{r}, \mathbf{p}) \to (g\mathbf{r}, \mathbf{p}g^T).$$

Mathematicians sometimes call this the diagonal action of the group $SO(3)$ on $\mathbb{R}^6$. We will call it the *rotation action of $SO(3)$* (on $\mathbb{R}^3 \times (\mathbb{R}^3)^*$).

Let us show that $S_g^* \omega = \omega$. Remember, the matrix g is constant, so we can pull it in and out of the exterior derivative d and the wedge product

$$d(\mathbf{pg}^T) \wedge d(\mathbf{gr}) = (d\mathbf{pg}^T) \wedge (\mathbf{g}d\mathbf{r}) = d\mathbf{p} \wedge (\mathbf{g}^T\mathbf{g})d\mathbf{r} = d\mathbf{p} \wedge d\mathbf{r},$$

since g is an orthogonal matrix and hence $g^T g = I$. So $S_g^* \omega = \omega$.

Let us redo this calculation in coordinates for two reasons. First, readers who are not yet comfortable manipulating the exterior product, the wedge and matrices all at once may find it helpful; second, the exercise should convince any skeptics that the coordinate-free notation is worth mastering.

$$d(\mathbf{pg}^T) \wedge d(\mathbf{gr}) = \begin{pmatrix} dp_x & dp_y & dp_z \end{pmatrix} \begin{pmatrix} g_{xx} & g_{yx} & g_{zx} \\ g_{xy} & g_{yy} & g_{zy} \\ g_{xz} & g_{yz} & g_{zz} \end{pmatrix} \wedge$$

$$\begin{pmatrix} g_{xx} & g_{xy} & g_{xz} \\ g_{yx} & g_{yy} & g_{yz} \\ g_{zx} & g_{zy} & g_{zz} \end{pmatrix} \begin{pmatrix} dr_x \\ dr_y \\ dr_z \end{pmatrix}$$

$$= (g_{xx}dp_x + g_{xy}dp_y + g_{xz}dp_z) \wedge (g_{xx}dr_x + g_{xy}dr_y + g_{xz}dr_z)$$
$$+ (g_{yx}dp_x + g_{yy}dp_y + g_{yz}dp_z) \wedge (g_{yx}dr_x + g_{yy}dr_y + g_{yz}dr_z)$$
$$+ (g_{zx}dp_x + g_{zy}dp_y + g_{zz}dp_z) \wedge (g_{zx}dr_x + g_{zy}dr_y + g_{zz}dr_z)$$

$$= (g_{xx}^2 + g_{yx}^2 + g_{zx}^2)dp_x \wedge dp_r + (g_{xx}g_{xy} + g_{yx}g_{yy} + g_{zx}g_{zy})dp_x \wedge dr_y$$
$$+ \text{seven more terms}$$

$$= 1 \, dp_x \wedge dr_x + 0 \, dp_x \wedge dr_y + \text{seven more terms}$$
$$= dp_x \wedge dr_x + dp_y \wedge dr_y + dp_\wedge dr_z$$
$$= d\mathbf{p} \wedge d\mathbf{r},$$

where the third equality follows because the columns of g are orthonormal. Which calculation do you prefer? In coordinates or coordinate-free?

Now let us check invariance of H, that is, check that $H \circ S_g = H$.

$$H \circ S_g(\mathbf{r}, \mathbf{p}) = H(\mathbf{gr}, \mathbf{pg}^T) = \frac{1}{2m} |\mathbf{pg}^T|^2$$

$$= \frac{1}{2m}\mathbf{pg}^T\mathbf{gp}^T = \frac{1}{2m}\mathbf{pp}^T = \frac{1}{2m}\mathbf{p}^2 = H(\mathbf{r}, \mathbf{p}).$$

To summarize in mathematical language, we have an action of the Lie group $SO(3)$ on $(\mathbb{R}^6, d\mathbf{p} \wedge d\mathbf{r}, \frac{1}{2m}|\mathbf{p}|^2)$ preserving both the symplectic form and the

Hamiltonian. We will see in Chapter 7 that this symmetry is related to conservation of angular momentum.

Note that all of the symmetries we have seen are *linear* symmetries, i.e., they can be expressed in matrix form. For the two-body problem it suffices to consider this kind of symmetry, but in other systems nonlinear symmetries often arise. For example, the time flow of the pendulum with gravity is not a linear symmetry of the system.

Exercise 96 *Check your fluency in coordinate-free calculations by deriving the Hamiltonian vector field for the flow of the two-body problem on the phase space* $\mathbb{R}^{12}$ *with symplectic form* $\omega = d\mathbf{p}_1 \wedge d\mathbf{r}_1 + d\mathbf{p}_2 \wedge d\mathbf{r}_2$. *The Hamiltonian function is*

$$H = \frac{1}{2}\left(\frac{|\mathbf{p}_1|^2}{m_1} + \frac{|\mathbf{p}_2|^2}{m_2}\right) - \frac{Gm_1m_2}{|\mathbf{r}_1 - \mathbf{r}_2|}.$$

Use Equation 4.4 to derive

$$X_H = \frac{\mathbf{p}_1}{m_1}(\frac{\partial}{\partial \mathbf{r}_1})^T + \frac{\mathbf{p}_2}{m_2}(\frac{\partial}{\partial \mathbf{r}_2})^T + \frac{Gm_1m_2}{|\mathbf{r}_1 - \mathbf{r}_2|^3}\left(\frac{\partial}{\partial \mathbf{p}_2} - \frac{\partial}{\partial \mathbf{p}_1}\right)(\mathbf{r}_1 - \mathbf{r}_2).$$

6
Infinitesimal Symmetries are Lie Algebras

In this chapter we introduce Lie algebras and show how a Lie group action on a manifold determines a vector field on the manifold for each element of the associated Lie algebra.

Mathematically, one can think of a Lie algebra $\mathfrak{g}$ as the set of possible derivatives of parametrized paths at the unit element 1 of a Lie group G. In other words, consider any parametrized path $g(t)$ in the group that passes through the identity element at time 0, i.e., with $g(0) = 1$. Then $g'(0)$ is an element of the Lie algebra $\mathfrak{g}$, and each element of the Lie algebra $\mathfrak{g}$ can be found as such a derivative. Examples of Lie algebras arise in the undergraduate physics curriculum, though the language barrier between mathematics and physics makes it hard to recognize them. One source of confusion is that physicists often take what mathematicians would call derivatives by working with infinitesimals. As an example, let us recover the product rule (also known as the Leibniz rule) by calculating with infinitesimals. Suppose x and y depend on time t. We write $x + dx$ for the value of x after a short time dt and $y + dy$ for the value of y after a short time dt. What is the value of xy after a short time dt? Since

$$(x + dx)(y + dy) = xy + (x \, dy + y \, dx) + \text{higher order terms}$$

we can conclude, by ignoring higher-order terms in the infinitesimals, that

$$\frac{d}{dt}(xy) = x\frac{dy}{dt} + y\frac{dx}{dt}.$$

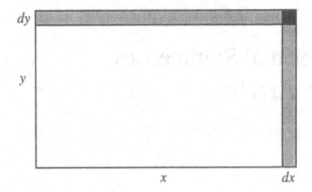

Figure 6.1. The product rule

See Figure 6.1. This is a perfectly correct procedure for calculating derivatives, and one can apply it to all kinds of objects. So one can think of Lie algebra elements as first-order terms in the Taylor series expansion of a group element g close to the identity. The elements of $\mathfrak{g}$ are sometimes called *infinitesimal symmetries* or *infinitesimal generators of the group G.*

6.1   Matrix Lie Algebras

We will start with matrix Lie algebras, which are all we need for our analysis of the two-body problem.

Definition 23 *A* matrix Lie algebra *is a vector space of matrices that is closed under the bracket operation* $[A, B] := AB - BA.$

Every matrix Lie group has an associated matrix Lie algebra.

Definition 24 *Given a matrix Lie group G, the* matrix Lie algebra $\mathfrak{g}$ of G *is the set of matrices*

$$\{g'(0) : g \text{ is a differentiable function from } \mathbb{R} \text{ to } G \text{ with } g(0) = I\},$$

where the derivative of a matrix-valued function is given by the matrix of derivatives of each coordinate entry.

It is important to prove that these definitions are consistent, i.e., that the matrix Lie algebra $\mathfrak{g}$ of a matrix Lie group G is indeed a vector space and is closed under the bracket operation. We postpone this proof to Section 6.3.

Exercise 97 *Find an invertible linear function F from $\mathbb{R}^{n\times n}$, the space of column $(n \times n)$-vectors, to the vector space of $n \times n$ matrices. Let g be a differ-*

entiable function from $\mathbb{R}$ to an $n \times n$ matrix Lie group G. Find an expression for g' in terms of the Jacobian of $F^{-1} \circ g$.

Let us apply the infinitesimal calculus to find a way to write infinitesimal rotations of three-space. In other words, let us find the Lie algebra of the Lie group $SO(3)$ introduced in Example 5.2. The mathematical name for this Lie algebra is "$so(3)$." Note that the letters are lower case, which distinguishes it on the page but, alas, not in speech, from the group $SO(3)$. Speakers conscientious enough to distinguish between "big-ess-oh-three" and "little-ess-oh-three" are admired and appreciated. Because $SO(3)$ is the *special orthogonal group* the Lie algebra $so(3)$ is known as the *special orthogonal algebra*, even though the matrices in $so(3)$ are neither orthogonal nor special (i.e., of determinant 1). (For a physical discussion of this Lie algebra, see the section entitled "Infinitesimal Rotations" in [Go].) We first calculate $so(3)$ using infinitesimals; then we will recalculate it in more formal mathematical language.

Recall that the Lie group of all rotations in three-space is

$$SO(3) = \{R : R \text{ is a real } 3 \times 3 \text{ matrix with } R^T R = I \text{ and } \det R = 1\}.$$

An infinitesimal rotation is something small you can add to the identity rotation without leaving the set of rotations. *Small* means "small enough that higher order terms can be ignored." Let ξ denote an arbitrary infinitesimal rotation. (The Greek letter ξ, pronounced "k-see" or "k-sigh" is often used to denote Lie algebra elements.) Then $(I + \xi)$ must be a rotation matrix, i.e., an element of $SO(3)$. By the definition of $SO(3)$, then, ξ is an infinitesimal rotation if and only if $(I + \xi)^T (I + \xi) = I$ and $\det(I + \xi) = 1$. Using the first equation and ignoring higher order terms in the infinitesimal ξ, we find that $I = I + \xi^T + \xi$, which is true precisely when ξ is a antisymmetric matrix. In other words, we have shown that any infinitesimal rotation can be written in coordinates as

$$\xi = \begin{pmatrix} 0 & -\xi_z & \xi_y \\ \xi_z & 0 & -\xi_x \\ -\xi_y & \xi_x & 0 \end{pmatrix}. \tag{6.1}$$

We will see in Exercise 100 why the notation is apt.

Next we must see which antisymmetric matrices satisfy the determinant condition. We require that

$$1 = \det \begin{pmatrix} 1 & -\xi_z & \xi_y \\ \xi_z & 1 & -\xi_x \\ -\xi_y & \xi_x & 1 \end{pmatrix} = 1 + \xi_x^2 + \xi_z^2 + \xi_y^2$$

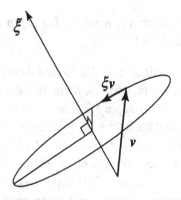

Figure 6.2.   The rotation vector

up to first order. But the right-hand side contains only the constant term 1 and terms of order two. So for any ξ this equation is satisfied. So the set of infinitesimal rotations is exactly the set of antisymmetric 3×3 matrices.

Exercise 98 *Show that for any infinitesimal matrix ξ (not necessarily antisymmetric), we have $\det(I+\xi) = 1$ if and only if $\operatorname{tr} \xi = 0$. Conclude that the matrix Lie algebra of the matrix Lie group $SL(n, \mathbb{R})$ is*

$$sl(n, \mathbb{R}) := \{\xi : \xi \text{ is an } n \times n \text{ real matrix and } \operatorname{tr} \xi = 0\}$$

Exercise 99 *Show that the matrix Lie algebra of the matrix Lie group $GL(n, \mathbb{R})$ is*

$$gl(n, \mathbb{R}) := \{\xi : \xi \text{ is an } n \times n \text{ real matrix}\}$$

To understand the matrix in Equation 6.1 geometrically as an infinitesimal rotation, it is convenient to introduce its *rotation vector*.

Definition 25 *Given a matrix ξ as in Equation 6.1, define its* rotation vector

$$\tilde{\xi} := \begin{pmatrix} \xi_x \\ \xi_y \\ \xi_z \end{pmatrix}.$$

See Figure 6.2. One implication of Exercise 100 is that the rotation vector $\tilde{\xi}$ is the axis of the infinitesimal rotation ξ.

Exercise 100 *Prove the following facts about the antisymmetric 3×3 matrix ξ and the column vector $\tilde{\xi}$:*

1. *For any vector $\mathbf{r}$ we have $\xi\mathbf{r} = \tilde{\xi} \times \mathbf{r}$.*

2. The vector $\tilde{\xi}$ is an eigenvector for ξ with eigenvalue 0.

3. If a vector $\mathbf{v}$ is perpendicular to $\tilde{\xi}$, then the vector $\xi\mathbf{v}$ is also perpendicular to $\tilde{\xi}$. In other words, the plane of vectors perpendicular to $\tilde{\xi}$ is an invariant space for the matrix ξ. Call this plane $\tilde{\xi}^T$.

4. If we restrict the linear transformation ξ to the plane $\tilde{\xi}^T$, we get an infinitesimal rotation of that plane, and any infinitesimal rotation of any plane comes from some antisymmetric matrix.

Exercise 101 *Show that the bracket of any two antisymmetric 3×3 matrices ξ and η is related to the cross-product of the axes of rotation by*

$$\widetilde{[\xi, \eta]} = -\tilde{\xi} \times \tilde{\eta}.$$

Physics students are familiar with infinitesimal rotations in the guise of angular velocity vectors. Physicists often denote angular velocity vector by ω and make great use of the formula $\frac{d}{dt}\mathbf{e} = \omega \times \mathbf{e}$ for any unit vector $\mathbf{e}$ that rotates with angular velocity ω. For example, see Baierlein [Bai, Chapter 6]. The angular velocity ω is defined to point along the axis of rotation, while the length of ω is the velocity of a point one unit distant from the rotation axis.

Exercise 102 *Show that the length $|\omega|$ of the angular velocity vector is equal to 2π divided by the period of rotation.*

Exercise 103 *Prove that $\frac{d}{dt}\mathbf{e} = \omega \times \mathbf{e}$ for any unit vector $\mathbf{e}$ with rotates with angular velocity ω.*

Exercise 104 *Suppose that ω is the angular velocity of a rotating system. Suppose that ξ is an antisymmetric matrix with $\tilde{\xi} = \omega$. Show that ξ is the linear operator taking each position vector to its velocity. In other words, show that for all vectors $\mathbf{r} \in \mathbb{R}^3$ we have $\frac{d}{dt}\mathbf{r} = \xi\mathbf{r}$.*

Next we recalculate $so(3)$ in mathematical language. We find the Lie algebra $so(3)$ of the group $SO(3)$ by differentiating parametrized paths through the identity element I. Suppose that $R : \mathbb{R} \to SO(3)$ is a differentiable function such that $R(0) = I$. Then $R'(0)$ is an element of the Lie algebra (which is the same as saying that $R'(0)$ is an infinitesimal rotation). Every element of the Lie algebra $so(3)$ arises this way. We will prove that

$$so(3) = \{\xi : \xi \text{ is a } 3 \times 3 \text{ real matrix and } \xi^T = -\xi\}. \tag{6.2}$$

First we show that the Lie algebra is contained in the set of antisymmetric matrices. Consider R as above. Since for every $s \in \mathbb{R}$ the matrix $R(s)$ lies in $SO(3)$, we have $R(s)^T R(s) = I$. Differentiating both sides of this equation and setting $s = 0$ we find that $R'(0)^T + R'(0) = 0$. In other words, if $R : \mathbb{R} \to SO(3)$ is any parametrized path in $SO(3)$ with $R(0) = I$ then $R'(0)$ is a antisymmetric matrix. So by Definition 23 any element of $so(3)$ is an antisymmetric matrix.

Next we show that the set of antisymmetric matrices is contained in the Lie algebra. We will use matrix exponentiation; see Section 0.2 for references. Let ξ be an arbitrary 3×3 antisymmetric matrix. We must find an $SO(3)$-valued function R such that $R'(0) = \xi$. Define $R : \mathbb{R} \to SO(3)$ by

$$R(s) := e^{s\xi}.$$

Notice first that for any s we have $R(s) \in SO(3)$ since

$$(e^{s\xi})^T (e^{s\xi}) = e^{s\xi^T} e^{s\xi} = e^{-s\xi} e^{s\xi} = I$$

and also $\det e^{s\xi} = 1$ as a consequence of Exercise 98. Then check that the derivative of R at $s = 0$ is indeed ξ. This ends the proof of Equation 6.2.   ◊

Exercise 105 *Check each step of the proof of Equation 6.2.*

The reader may wish to compare the two calculations of $so(3)$. How did the first (physics-style) avoid exponentiation of matrices?

Next we find the Lie algebra of the translation symmetry group for the two-body problem (introduced in Example 5.2). The group is $(\mathbb{R}^3, +)$. Let us find its Lie algebra by the infinitesimal calculus. Note that the identity element of this group is $\mathbf{0}$. If $\boldsymbol{\xi}$ is an infinitesimal translation, then $\mathbf{0} + \boldsymbol{\xi} \in \mathbb{R}^3$, so $\boldsymbol{\xi} \in \mathbb{R}^3$. In other words, the Lie algebra of the group $(\mathbb{R}^3, +)$ is the vector space $\mathbb{R}^3$.

For readers who are sticklers for uniform mathematical calculations, let us find the Lie algebra of the group $(R^3, +)$ again, this time by exhibiting $(R^3, +)$ as a matrix group and proceeding as we did for $SO(3)$.

Recall from Section 5.2 that the group $(\mathbb{R}^3, +)$ is isomorphic to the matrix group

$$G := \left\{ \begin{pmatrix} 1 & 0 & 0 & g_x \\ 0 & 1 & 0 & g_y \\ 0 & 0 & 1 & g_z \\ 0 & 0 & 0 & 1 \end{pmatrix} : g_x, g_y, g_z \in \mathbb{R} \right\}.$$

Differentiable parametrized paths through the identity element inside this group have the form

$$g(s) = \begin{pmatrix} 1 & 0 & 0 & g_x(s) \\ 0 & 1 & 0 & g_y(s) \\ 0 & 0 & 1 & g_z(s) \\ 0 & 0 & 0 & 1 \end{pmatrix},$$

where g_x, g_y and g_z are arbitrary differentiable functions from $\mathbb{R}$ to $\mathbb{R}$ and $g_1(0) = g_2(0) = g_3(0) = 0$. Differentiating entry by entry yields:

$$g'(0) = \begin{pmatrix} 0 & 0 & 0 & g'_x(0) \\ 0 & 0 & 0 & g'_y(0) \\ 0 & 0 & 0 & g'_z(0) \\ 0 & 0 & 0 & 0 \end{pmatrix}.$$

We leave it to the reader to show that every matrix of this form is in the matrix Lie algebra of the matrix Lie group G:

Exercise 106 *For each element ξ of $\mathbb{R}^3$, construct a parametrized path g through the identity in the matrix Lie group G so that*

$$g'(0) = \begin{pmatrix} 0 & 0 & 0 & \xi_x \\ 0 & 0 & 0 & \xi_y \\ 0 & 0 & 0 & \xi_z \\ 0 & 0 & 0 & 0 \end{pmatrix}.$$

Thus the matrix Lie algebra of G is

$$\mathfrak{g} = \left\{ \begin{pmatrix} 0 & 0 & 0 & \xi_x \\ 0 & 0 & 0 & \xi_y \\ 0 & 0 & 0 & \xi_z \\ 0 & 0 & 0 & 0 \end{pmatrix} : \xi_x, \xi_y, \xi_z \in \mathbb{R} \right\}.$$

This matrix Lie algebra is called *commutative*, or *abelian* because for all A, $B \in \mathfrak{g}$ we have $[A, B] = 0$, as the reader can easily check by matrix multiplication.

6.2 Vector Fields Associated to Lie Algebra Elements

To understand the momentum map (which will be introduced in Chapter 7) we must define the vector fields associated to Lie algebra elements via a Lie group action. The first step is to understand one-parameter subgroups. Although one-parameter subgroups can be defined for any Lie group (see, e.g., [AM, p. 255]), we will limit ourselves to one-parameter subgroups of matrix Lie groups for simplicity's sake.

Definition 26 *Given a matrix Lie group G and an element ξ of its Lie algebra, the one-parameter subgroup generated by ξ is*

$$G_\xi := \left\{ e^{s\xi} : s \in \mathbb{R} \right\}.$$

Let $G = SO(3)$ and consider

$$\xi = \begin{pmatrix} 0 & -1 & 0 \\ 1 & 0 & 0 \\ 0 & 0 & 0 \end{pmatrix} \in so(3).$$

Then for any $s \in \mathbb{R}$, by the definition of matrix exponentiation,

$$e^{s\xi} = \begin{pmatrix} \cos s & -\sin s & 0 \\ \sin s & \cos s & 0 \\ 0 & 0 & 1 \end{pmatrix},$$

as one can check either by diagonalizing ξ (over the complex numbers) or by calculating the power series expansion

$$e^{s\xi} = I + s\xi + \frac{1}{2!}(s\xi)^2 + \frac{1}{3!}(s\xi)^3 + \cdots.$$

In this case the group G_ξ is the group of rotations around the axis $\tilde{\xi} = (0, 0, 1)^T$ obeying the right-hand rule. (That is, if you point your right thumb in the $\tilde{\xi}$-direction, your right fingers will curl in the direction of increasing s.) So $G_\xi \cong S^1$.

Exercise 107 *For which $\xi \in so(3)$ is $G_\xi \cong S^1$?*

Let us find the one-parameter subgroups of $(\mathbb{R}^3, +)$. Let ξ be an arbitrary element of the Lie algebra $\mathbb{R}^3$ of $(\mathbb{R}^3, +)$. Because our definition is for matrix Lie groups, we must work with the matrix Lie group form of this group (see Example 5.2). Accordingly, we write

$$\xi = \begin{pmatrix} 0 & 0 & 0 & \xi_x \\ 0 & 0 & 0 & \xi_y \\ 0 & 0 & 0 & \xi_z \\ 0 & 0 & 0 & 0 \end{pmatrix}$$

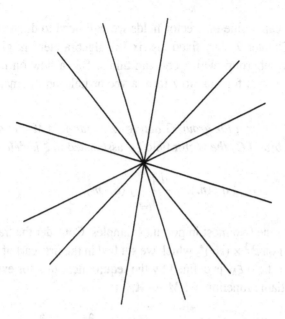

Figure 6.3. The one-parameter subgroups of $(\mathbb{R}^2, +)$

and find that for any $s \in \mathbb{R}$ we have

$$
e^{s\xi} = \begin{pmatrix} 1 & 0 & 0 & s\xi_x \\ 0 & 1 & 0 & s\xi_y \\ 0 & 0 & 1 & s\xi_z \\ 0 & 0 & 0 & 1 \end{pmatrix}.
$$

The corresponding element of $(\mathbb{R}^3, +)$ is $s\xi = (s\xi_x, s\xi_y, s\xi_z)^T$. So the one-parameter subgroup of $(\mathbb{R}^3, +)$ associated to $\xi \in \mathbb{R}^3$ is the line parallel to ξ through the origin in $(\mathbb{R}^3, +)$. See Figure 6.3 for the one-parameter subgroups of the two-dimensional group $(\mathbb{R}^2, +)$.

Exercise 108 *Consider the torus group $G = S^1 \times S^1$. We can write G concretely as*

$$
G = \left\{ \begin{pmatrix} \cos(s) & -\sin(s) & 0 & 0 \\ \sin(s) & \cos(s) & 0 & 0 \\ 0 & 0 & \cos(u) & -\sin(u) \\ 0 & 0 & \sin(u) & \cos(u) \end{pmatrix} : s, u \in \mathbb{R} \right\}.
$$

Find its Lie algebra $\mathfrak{g}$. For which $\xi \in \mathfrak{g}$ is $G_\xi \cong S^1$? Extra credit: For which $\xi \in \mathfrak{g}$ is G_ξ a Lie subgroup of G?
You can find the definition of a *Lie subgroup* in [AM, Def. 4.1.10] or [Br, Lecture 2, Def. 2].

Finally, we can define the vector fields we will need to define the momentum map in Chapter 7. Any fixed matrix Lie algebra element gives rise to a one-parameter subgroup, which one can think of as a flow on the manifold. The velocity vectors for this flow form a vector field on the manifold. More precisely:

Definition 27 *Given a Lie group G acting on a manifold M and an element ξ of the Lie algebra of G, the* vector field ξ_M associated to ξ *is defined by*

$$\xi_M f(m) = \left.\frac{d}{dt}\right|_{t=0} f(S_{e^{t\xi}} m).$$

We calculate the two most important examples. Consider the translation action of $(\mathbb{R}^3, +)$ on $\mathbb{R}^3 \times (\mathbb{R}^3)^*$, which we studied in the first part of Section 5.6. Here the vector field ξ_M is defined by the requirement that for every continuously differentiable function $f : M \to \mathbb{R}$,

$$\xi_M f(\mathbf{r}, \mathbf{p}) := \left.\frac{d}{dt}\right|_{t=0} f(\mathbf{r} + t\xi, \mathbf{p}) = \frac{\partial f}{\partial r_x}\xi_x + \frac{\partial f}{\partial r_y}\xi_y + \frac{\partial f}{\partial r_z}\xi_z = \frac{\partial f}{\partial \mathbf{r}}\xi.$$

So $\xi_M = \frac{\partial}{\partial \mathbf{r}}\xi = \xi_x\frac{\partial}{\partial r_x} + \xi_y\frac{\partial}{\partial r_y} + \xi_z\frac{\partial}{\partial r_z}$. Geometrically, the vector field ξ_M assigns the tangent vector $(\xi, \mathbf{0})$ to each point of the manifold $M = \mathbb{R}^3 \times (\mathbb{R}^3)^*$. This vector field is sometimes called the *infinitesimal generator of translation in the ξ-direction*.

Some of the notation used in the previous paragraph may be unfamiliar to the reader. Here $\frac{\partial f}{\partial \mathbf{r}} = (\frac{\partial f}{\partial r_x}, \frac{\partial f}{\partial r_y}, \frac{\partial f}{\partial r_z})$ is a row vector, and $\frac{\partial}{\partial \mathbf{r}}$ is a row vector as well:

$$\frac{\partial}{\partial \mathbf{r}} = \left(\begin{array}{ccc} \frac{\partial}{\partial r_x} & \frac{\partial}{\partial r_y} & \frac{\partial}{\partial r_z} \end{array} \right).$$

Because $\mathbf{p}$ is a row vector, $\frac{\partial f}{\partial \mathbf{p}}$ and $\frac{\partial}{\partial \mathbf{p}}$ should be column vectors. Here our notation becomes slightly inconvenient, as the rules of matrix multiplication dictate that $\xi_x\frac{\partial}{\partial r_x} + \xi_y\frac{\partial}{\partial r_y} + \xi_z\frac{\partial}{\partial r_z}$ should be written

$$\left(\begin{array}{c} \frac{\partial}{\partial \mathbf{r}} \end{array} \right) \left(\begin{array}{c} \xi \end{array} \right)$$

but the conventions of differentiation would suggest that we should differentiate ξ with respect to $\mathbf{r}$, which we do not want to do. We will respect the matrix multiplication, and write $\frac{\partial}{\partial \mathbf{r}}\xi$, asking the reader to remember when and when not to differentiate. This notational tension can be resolved by using the tensor index notation and Einstein summation convention popular with physicists. See, for example, [Go, Sections 4.2 and 7.3].

Next we consider the rotation action of the group $SO(3)$ on the manifold $\mathbb{R}^3 \times (\mathbb{R}^3)^*$. In this context an element ξ of the Lie algebra $so(3)$ is an antisymmetric 3×3 matrix. From the definition of ξ_M we have, for every differentiable function $f : M \to \mathbb{R}$,

$$\xi_M f(\mathbf{r}, \mathbf{p}) = \frac{d}{dt}\bigg|_{t=0} f(e^{t\xi}\mathbf{r}, \mathbf{p}e^{t\xi^T}) = \frac{\partial f}{\partial \mathbf{r}}\xi\mathbf{r} + \mathbf{p}\xi^T\frac{\partial f}{\partial \mathbf{p}}$$

$$\tag{6.3}$$

$$= \left(\frac{\partial}{\partial \mathbf{r}}\xi\mathbf{r} - \mathbf{p}\xi\frac{\partial}{\partial \mathbf{p}}\right) f,$$

where we use the notation convention described above. So $\xi_M = \frac{\partial}{\partial \mathbf{r}}\xi\mathbf{r} - \mathbf{p}\xi\frac{\partial}{\partial \mathbf{p}}$. We encourage the reader to check the dimensions of the matrices and vectors on the right-hand side of this equation to verify that it is indeed a 1×1 matrix whose entry is a single first-order linear differential operator. This vector field is *the infinitesimal generator of rotations around the $\tilde{\xi}$ axis.*

Exercise 109 *Use the chain rule and Jacobian matrices to check the second equality in Equation 6.3.*

6.3   Abstract Lie Algebras (Optional)

Abstract Lie algebras are mathematical objects in their own right, with enough structure to be studied without reference to Lie groups. We provide a brief introduction for the interested reader.

Definition 28 *A Lie algebra $\mathfrak{g}$ is a vector space equipped with a bracket operation*

$$[,] : \mathfrak{g} \times \mathfrak{g} \to \mathfrak{g}$$

that is bilinear, antisymmetric and satisfies the Jacobi identity

$$[[A, B], C] + [[B, C], A] + [[C, A], B] = 0.$$

This bracket is called the Lie bracket.

Any matrix Lie algebra (see Definition 23) is indeed a Lie algebra, since the bracket $[A, B] := AB - BA$ is bilinear and antisymmetric and $[[A, B], C] + [[B, C], A] + [[C, A], B] = (AB - BA)C - C(AB - BA) + (BC - CB)A - A(BC - CB) + (CA - AC)B - B(CA - AC) = 0$.

For any matrix Lie group G the corresponding matrix Lie algebra $\mathfrak{g}$ (see Definition 24) is a Lie algebra. By the argument in the previous paragraph, it suffices to show that $\mathfrak{g}$ is a vector space of matrices closed under the bracket operation. The fact that $\mathfrak{g}$ is closed under addition follows from the product rule for differentiating matrices: if g_A and g_B are parametrized paths in the group with $g_A(0) = 1$, $g_A'(0) = A$, $g_B(0) = 1$ and $g_B'(0) = B$, then $g_A g_B$ is also a path through the identity in the group and $(g_A g_B)'(0) = g_A'(0) g_B(0) + g_A(0) g_B'(0) = A + B$. Closure under scalar multiplication follows from the chain rule: for any constant $c \in \mathbb{R}$ we have $\frac{d}{ds} g_A(cs) = c g_A'(cs)$ and hence $cA \in \mathfrak{g}$. Closure under the bracket requires a slightly longer argument. Given A and B in $\mathfrak{g}$, we consider the matrix-valued function $g_A(u) g_B(s) g_A(u)^{-1}$ of two real variables, u and s. For any fixed u this expression is a parametrized path through the identity in the group G, so it follows that for any u, we have $g_A(u) B g_A(u)^{-1}$ in $\mathfrak{g}$. But $\mathfrak{g}$ is a finite-dimensional vector space, so if we differentiate this last expression with respect to u, the result will also lie in $\mathfrak{g}$. Evaluating the derivative at $u = 0$, we find $AB - BA \in \mathfrak{g}$.

Note that if the matrix Lie group is commutative, the bracket is trivial. In general, one may think of the Lie bracket as measuring (to some extent) the noncommutativity of the corresponding Lie group. See [Ar89, Section 39].

Exercise 110 *Check that if $A \in so(3)$ and $B \in so(3)$, then $[A, B] \in so(3)$.*

Every Lie group (defined in Section 5.3) has an associated Lie algebra. The *Lie algebra $\mathfrak{g}$ of G* is the tangent space to the group G at the identity element 1, endowed with a certain naturally defined Lie bracket operation. Defining the bracket in terms of the group and manifold structure of the Lie group requires the notion of left-invariant vector fields, which is too much of a detour for this text. We recommend the treatment in Bröcker and tom Dieck [BtD, Section I.2]. Readers who value brevity should consult Abraham and Marsden [AM, Section 4.1].

7
Conserved Quantities
are Momentum Maps

Conserved quantities are powerful tools in physics, both conceptually and for calculations. The conceptual power of conserved quantities such as momentum and energy is evident from their acceptance into everyday speech. Equations expressing conservation abound in elementary physics courses because they allow one to make informative calculations without using the techniques of calculus.

A very general formulation of the relationship between symmetries and conserved quantities was given by Emmy Noether, who showed that associated to any one-dimensional symmetry group of a mechanical system there is a (possibly only locally defined) conserved quantity. This result is known as Noether's Theorem. The most natural setting for this theorem is in terms of Lagrangians and the calculus of variations; see [Ar89, Section 20] or [O, Sections 4.4 and 5.3]. In the Hamiltonian formulation of mechanics (the focus of this book) the relationship between symmetries and conserved quantities is expressed in terms of group actions and *momentum maps*. The momentum map for a particular group action is a function on phase space that generates the group action in a way we will make precise in Section 7.1. In Section 7.2 we will show that linear momentum and angular momentum, the conserved quantities that play the major role in the analysis of the two-body problem, are bona fide examples of momentum maps. In the optional Section 7.3 we use some sophisticated differential geometry to show how to construct momentum maps from group actions on mechanical phase spaces.

7.1   Definition of Momentum Map

In this section we generalize the notion of a *dual space* (introduced in Section 2.2) to arbitrary vector spaces. We are interested particularly in dual spaces of Lie algebras, which we use to define *momentum maps*.

Suppose V is a vector space. Denote by V^* the *dual space of the vector space V*, i.e., the set of linear functions from V to $\mathbb{R}$. Linear functions from V to $\mathbb{R}$ are often called *linear functionals on V*, so the dual V^* is the *vector space of linear functionals on V*.)

Exercise 111 *Check that V^* is a vector space. Specifically, use the natural addition and scalar multiplication of functions and check the relevant axioms.*

To understand any particular dual space concretely it helps to choose a way to write down its elements. In Section 2.2 we studied $(\mathbb{R}^2)^*$ and wrote its elements as row vectors. We used matrix multiplication to interpret row vectors as real-valued functions on column vectors. When the vector space in question is a space of matrices, it is often useful to consider the trace of a product of matrices, which provides a way to construct linear functionals from matrices.

Exercise 112 *Let V be a vector space of $n \times m$ matrices. Show that for each $m \times n$ matrix A, the function $V \to \mathbb{R}$, $B \mapsto \mathrm{tr}(AB)$ is a linear functional.*

For example, we can think of the dual space to the matrix Lie algebra

$$\mathfrak{g} := \left\{ \begin{pmatrix} 0 & 0 & 0 & \xi_x \\ 0 & 0 & 0 & \xi_y \\ 0 & 0 & 0 & \xi_z \\ 0 & 0 & 0 & 0 \end{pmatrix} : \xi_x, \xi_y, \xi_z \in \mathbb{R} \right\}$$

as the space of matrices

$$\mathfrak{g}^* = \left\{ \begin{pmatrix} 0 & 0 & 0 & 0 \\ 0 & 0 & 0 & 0 \\ 0 & 0 & 0 & 0 \\ \eta_x & \eta_y & \eta_z & 0 \end{pmatrix} : \eta_x, \eta_y, \eta_z \in \mathbb{R} \right\}.$$

Note that

$$
\text{tr}
\begin{pmatrix}
0 & 0 & 0 & 0 \\
0 & 0 & 0 & 0 \\
0 & 0 & 0 & 0 \\
\eta_x & \eta_y & \eta_z & 0
\end{pmatrix}
\begin{pmatrix}
0 & 0 & 0 & \xi_x \\
0 & 0 & 0 & \xi_y \\
0 & 0 & 0 & \xi_z \\
0 & 0 & 0 & 0
\end{pmatrix}
= \eta_x \xi_x + \eta_y \xi_y + \eta_z \xi_z,
$$

which is the most general form for a linear functional on $\mathfrak{g}$.

For the two-body problem we will need to write down elements of $so(3)^*$. We can write each element of $so(3)^*$ uniquely as an element of $so(3)$ itself by noting that, for any $\eta \in so(3)$, the function from $so(3)$ to $\mathbb{R}$ given by

$$
\xi \mapsto -\frac{1}{2}\text{tr}
\begin{pmatrix}
0 & -\eta_z & \eta_y \\
\eta_z & 0 & -\eta_x \\
-\eta_y & \eta_x & 0
\end{pmatrix}
\begin{pmatrix}
0 & -\xi_z & \xi_y \\
\xi_z & 0 & -\xi_x \\
-\xi_y & \xi_x & 0
\end{pmatrix}
\quad (7.1)
$$
$$
= (\eta_x \xi_x + \eta_y \xi_y + \eta_z \xi_z)
$$

is a linear functional on $so(3)$ and every linear functional on $so(3)$ has this form.

With the notion of the dual space of a Lie algebra in hand, we can give a definition of a momentum map. Fix a Lie group G. Call its Lie algebra $\mathfrak{g}$ and call the dual space $\mathfrak{g}^*$. Consider an infinitely differentiable function Φ from a symplectic manifold M to $\mathfrak{g}^*$. If for each $m \in M$ we write $(\Phi(m))(\xi)$ for the value of the linear functional $\Phi(m)$ on a Lie algebra element ξ, then we can define $\Phi^\xi : M \to \mathbb{R}$ for each $\xi \in \mathfrak{g}$ by

$$
\Phi^\xi := (\Phi(\cdot))(\xi).
$$

This function can be viewed as a Hamiltonian function. Hence it, along with the symplectic form on M, determines a Hamiltonian vector field X_{Φ^ξ}, as explained in Section 4.1. Thus the function Φ provides a way to associate a vector field to each Lie algebra element.

If the group G acts on M then, from Section 6.2, we have another way of associating a vector field to each Lie algebra element. Given any ξ in $\mathfrak{g}$, we can construct a vector field ξ_M by differentiating the group action, as explained in Definition 27. If, for each Lie algebra element ξ, the two constructions agree with one another, then we say that Φ is a momentum map for the action of G. Of course, two different elements ξ_1 and ξ_2 may generate two different vector fields, but for each individual ξ we should have $\xi_M = X_{\Phi^\xi}$. In other words:

Definition 29 *The function* $\Phi : M \rightarrow \mathfrak{g}^*$ *is a momentum map for the action of the group G on the symplectic manifold (M, ω) if and only if for every $\xi \in \mathfrak{g}$ we have*

$$\omega(\xi_M, \cdot) = -d\Phi^\xi(\cdot).$$

Whew! All that work for one definition! Time for some examples. But we pause to remark that if we fix a group G with Lie algebra $\mathfrak{g}$ it is *not* true that any map from M to $\mathfrak{g}^*$ is a momentum map for a G-action on M. Also, it is *not* true that every Lie group action has an associated momentum map. So *Hamiltonian actions of G*, i.e., group actions with associated momentum maps, are lovely special creatures.

As a first example, note that because $\mathbb{R}$ is itself the dual of the matrix Lie algebra of $GL(1, \mathbb{R})$, any Hamiltonian function H is a momentum map. Specifically, it is not hard to check that for each $\xi \in \mathbb{R}$ we have $H^\xi = \xi H$ and $\xi_M = \xi X_H$. So for every $\xi \in \mathbb{R}$ we have $\omega(\xi_M, \cdot) = \omega(\xi X_H, \cdot) = \xi\omega(X_H, \cdot) = \xi\iota_{X_H}\omega$. On the other hand, $-dH^\xi = -\xi dH$. But a Hamiltonian vector field must satisfy Equation 4.5 of Section 4.1. So $\omega(\xi_M, \cdot) = -dH^\xi$ and H is a momentum map. We work out two more examples in Section 7.2.

Explicit definitions of the momentum map appeared in print in various forms in 1970. Stephen Smale started from a symmetry group of a Hamiltonian system and constructed a map from the manifold to the dual of the Lie algebra without mentioning symplectic forms [Sm, Section 4]. Jean-Marie Souriau gave the definition we use in this book [So, Definition 11.7b]. Independently, Bertram Kostant gave an equivalent definition in [Kos, p. 172]. These and other researchers were involved in the development of the idea during the 1960s. For a more detailed history of the development of the idea during the 1960s, see Marsden and Ratiu [MR, Section 11.2].

The reader should be aware that the mathematical world is not united on some of the terminology used here. In Guillemin and Sternberg's work [GS84] the phrase "Hamiltonian group action" means an action having a corresponding momentum map. In Arnold and Givental's article [AG] the phrase "Hamiltonian group action" refers to a group action that preserves the symplectic form. Although every group action having a momentum map preserves the symplectic form, the converse is not true: try rotating a torus (see Exercise 114). That is, not every group action preserving the symplectic form has a momentum map.

7.2 Examples from the Two-Body Problem

In Chapter 5 we looked at certain Lie group actions of $(\mathbb{R}^3, +)$ and $SO(3)$. We will show now that these are both Hamiltonian actions, i.e., that they have associated momentum maps. The values of these momentum maps will turn out to be linear and angular momentum, which is the etymology of the name "momentum map."

Note that for this section we can forget about the Hamiltonian function H that determined the differential equations for the motion of the mechanical system. We considered it in Section 5.6, when we combined symmetries and energy flows, and will again in Chapter 8, when we do reduction. But for the next few pages it is perhaps easier to forget the Hamiltonian completely.

The reader may find it helpful at this time to review the remarks in Section 6.2 on our notation convention for vectors whose entries are differential operators.

For our first example we consider the translation action of $(\mathbb{R}^3, +)$ on $M = \{(\mathbf{r}_1, \mathbf{r}_2, \mathbf{p}_1, \mathbf{p}_2)\} = \mathbb{R}^3 \times \mathbb{R}^3 \times (\mathbb{R}^3)^* \times (\mathbb{R}^3)^*$, with symplectic form $\omega = d\mathbf{p}_1 \wedge d\mathbf{r}_1 + d\mathbf{p}_2 \wedge d\mathbf{r}_2$. This is the phase space of two particles moving in $\mathbb{R}^3$. We will show that the momentum map for this action maps the position and momentum data to the total linear momentum

$$\Phi : \qquad M \to (\mathbb{R}^3)^*$$
$$(\mathbf{r}_1, \mathbf{r}_2, \mathbf{p}_1, \mathbf{p}_2) \mapsto \mathbf{p}_1 + \mathbf{p}_2.$$

So $\Phi^{\xi}(\mathbf{r}_1, \mathbf{r}_2, \mathbf{p}_1, \mathbf{p}_2) = (\mathbf{p}_1 + \mathbf{p}_2)\xi$. Recall that $(\mathbb{R}^3)^*$ is the space of linear functions from $\mathbb{R}^3$ to $\mathbb{R}$. Specifically, for $i = 1$ or 2, the function denoted $\mathbf{p}_i$ is this one:

$$\mathbf{p}_i : \quad \mathbb{R}^3 \to \mathbb{R}$$
$$\xi \mapsto \mathbf{p}_i \xi,$$

where $\mathbf{p}_i \xi$ denotes matrix multiplication. In coordinates, if $\mathbf{p}_i = (p_{ix}, p_{iy}, p_{iz})$, then

$$\mathbf{p}_i(\xi) = \begin{pmatrix} p_{ix} & p_{iy} & p_{iz} \end{pmatrix} \begin{pmatrix} \xi_x \\ \xi_y \\ \xi_z \end{pmatrix} = p_{ix}\xi_x + p_{iy}\xi_y + p_{iz}\xi_z.$$

There is an action of $(\mathbb{R}^3, +)$ on M given by

$$S_g : \qquad M \to M$$
$$(\mathbf{r}_1, \mathbf{r}_2, \mathbf{p}_1, \mathbf{p}_2) \mapsto (\mathbf{r}_1 + \mathbf{g}, \mathbf{r}_2 + \mathbf{g}, \mathbf{p}_1, \mathbf{p}_2)$$

for each $\mathbf{g} \in \mathbb{R}^3$. This symmetry corresponds to two observers whose coordinate system origins are separated by the vector $\mathbf{g}$. It is not hard to calculate (following a calculation in Section 6.2) that

$$\xi_M = (\frac{\partial}{\partial \mathbf{r}_1} + \frac{\partial}{\partial \mathbf{r}_2})\xi$$

$$= \xi_x(\frac{\partial}{\partial r_{1x}} + \frac{\partial}{\partial r_{2x}}) + \xi_y(\frac{\partial}{\partial r_{1y}} + \frac{\partial}{\partial r_{2y}}) + \xi_z(\frac{\partial}{\partial r_{1z}} + \frac{\partial}{\partial r_{2z}}).$$

We now verify the definition of the momentum map in this example. We can plug our formula for Φ^ξ and our formula for ξ_M into Definition 29. For any fixed $\xi \in \mathbb{R}^3$ we have

$$d\Phi^\xi = d\left((\mathbf{p}_1 + \mathbf{p}_2)\xi\right) = (d\mathbf{p}_1 + d\mathbf{p}_2)\,\xi.$$

On the other hand,

$$\omega(\xi_M, \cdot) = (d\mathbf{p}_1 \wedge d\mathbf{r}_1 + d\mathbf{p}_2 \wedge d\mathbf{r}_2)((\frac{\partial}{\partial \mathbf{r}_1} + \frac{\partial}{\partial \mathbf{r}_2})\xi, \cdot) = -(d\mathbf{p}_1 + d\mathbf{p}_2)\,\xi.$$

So $-d\Phi^\xi(\cdot) = \omega(\xi_M, \cdot)$, and hence Φ is a momentum map for this action. Note that the value of the momentum map is the total linear momentum of the two particles.

For our second example we consider the rotation action of $SO(3)$ on

$$M = \{(\mathbf{r}, \mathbf{p})\} = \mathbb{R}^3 \times (\mathbb{R}^3)^* \cong \mathbb{R}^6$$
$$\omega = d\mathbf{p} \wedge d\mathbf{r}$$

with momentum map

$$\Phi: \qquad M \to (so(3))^*$$
$$(\mathbf{r}, \mathbf{p}) \mapsto -\left((\mathbf{rp})^T - \mathbf{rp}\right).$$

See Section 5.6. Note that if we think of $\mathbf{r}$ as a column vector and $\mathbf{p}$ as a row vector, then by matrix multiplication $-((\mathbf{rp})^T - \mathbf{rp})$ is a 3×3 antisymmetric matrix. Formula 7.1 tells us how to interpret this antisymmetric matrix as a linear functional on $so(3)$. So the linear function in $so(3)^*$ which is the value of the momentum map Φ at $(\mathbf{r}, \mathbf{p})$ is given explicitly by

$$\Phi(\mathbf{r}, \mathbf{p}): \qquad so(3) \to \mathbb{R}$$
$$\xi \mapsto -\tfrac{1}{2}\mathrm{tr}\left(-((\mathbf{rp})^T - \mathbf{rp})\xi\right).$$

We are not far now from the physicists' angular momentum vector $\tilde{\mathbf{L}} :=$ $\mathbf{r} \times \mathbf{p}^T$ from Chapter 1. In fact, the rotation vector of $-((\mathbf{rp})^T - \mathbf{rp})$ is precisely $\tilde{\mathbf{L}}$, according to Definition 25 of Section 6.1). So it is consistent to define $L :=$ $-((\mathbf{rp})^T - \mathbf{rp})$ and

$$\Phi^\xi(\mathbf{r}, \mathbf{p}) = -\frac{1}{2}\text{tr}(L\xi) = \tilde{\mathbf{L}}^T\tilde{\xi}. \tag{7.2}$$

Let us verify Definition 29, the definition of the momentum map, in this example. We will use the fact that $\xi^T = -\xi$ as well as some properties of the trace given in Exercise 2. Recall that the rotation action of $SO(3)$ on $\dot{M}$ is given by

$$S_g : \qquad M \to M$$

$$(\mathbf{r}, \mathbf{p}) \mapsto (g\mathbf{r}, \mathbf{p}g^T)$$

for any $g \in SO(3)$. We showed in Section 5.6 that this action is symplectic. Using the notation convention and results of Section 6.2 we have $\xi_M = \frac{\partial}{\partial \mathbf{r}}\xi\mathbf{r} - \mathbf{p}\xi\frac{\partial}{\partial \mathbf{p}}$. So for any fixed $\xi \in so(3)$ we have

$$d\Phi^\xi = d(-\frac{1}{2}\text{tr}\left(-(\mathbf{rp} - (\mathbf{rp})^T)\xi\right) = \frac{1}{2}\text{tr}(\mathbf{p}\xi d\mathbf{r} - \mathbf{p}\xi^T d\mathbf{r} + \xi \mathbf{r}d\mathbf{p} - \xi^T \mathbf{r}d\mathbf{p})$$

$$= \mathbf{p}\xi d\mathbf{r} + d\mathbf{p}\xi\mathbf{r},$$

while

$$\omega(\xi_M, \cdot) = d\mathbf{p} \wedge d\mathbf{r}(-\mathbf{p}\xi^T\frac{\partial}{\partial \mathbf{p}} + \frac{\partial}{\partial \mathbf{r}}\xi\mathbf{r}, \cdot) = -\mathbf{p}\xi d\mathbf{r} - d\mathbf{p}\xi\mathbf{r}.$$

We have shown that Φ is a momentum map for the rotation action of $SO(3)$. Note that the value of Φ is precisely the angular momentum of the particle with position $\mathbf{r}$ and momentum $\mathbf{p}$.

These two examples of group actions with momentum maps – translation with linear momentum and rotation with angular momentum – are the symmetries exploited in the analysis of the two-body problem. The reader may with to review Chapter 1 to see where each of these symmetries is used in the physics calculation.

7.3   Momentum Maps from Group Actions (Optional)

Given a particular Lie group action preserving a particular symplectic form, it is natural to look for a momentum map for that action. This comes down to solving a partial differential equation: we can calculate ξ_M from the given group action and use the given symplectic form to find the one-form $\iota_{\xi_M}\omega$. Then we look for a function $\Phi : M \to \mathfrak{g}^*$ satisfying $d\Phi^\xi = -\iota_{\xi_M}\omega$. More explicitly, if we write $\iota_{\xi_M}\omega = f_r dr + f_p dp$ we must solve the system of equations

$$\frac{\partial \Phi^\xi}{\partial r} = -f_r \text{ and } \frac{\partial \Phi^\xi}{\partial p} = -f_p$$

for the function Φ.

A more sophisticated mathematical question is whether every group action has an associated momentum map. The answer is often "no" (see for example Exercise 114) but "yes" in the special cases where the symplectic manifold is a mechanical phase space and the group action comes from coordinate changes on configuration space. (Recall that the configuration space is the set of all possible positions of the system, without considering velocity or momentum.) The concepts of differential geometry allow us to restate the condition more precisely: if M is the cotangent bundle of a manifold N (i.e., if M is the phase space and N is the configuration space), if ω is the natural symplectic form on this cotangent bundle, and if the group action on the bundle manifold M is the lift of a group action on the base manifold N, then the group action has a momentum map. This was first published in 1970, by Smale [Sm].

For readers comfortable with differential geometry (including cotangent bundles and Lie derivatives, see Warner [Wa, Chapters 1 and 2]) we sketch the construction of the momentum map from the group action in this case. The key ingredient is a distinguished one-form β on M which satisfies:

1. $d\beta = \omega$;

2. β is invariant under any group action on M lifted from an action on the base manifold N.

Such a β, known as the *canonical one-form*, can be constructed on any cotangent bundle; for a proof, see Abraham and Marsden [AM, Proposition 4.1.10]. We will content ourselves with working out two examples.

For example, if $N = \mathbb{R}^3 \times \mathbb{R}^3 = \{r_1, r_2\}$ and $M = \mathbb{R}^{12} = \{(r_1, r_2, p_1, p_2)\}$ with $\omega = dp_1 \wedge dr_1 + dp_2 \wedge dr_2$, then we can take $\beta = p_1 dr_1 + p_2 dr_2$. Then $d\beta = dp_1 \wedge dr_1 + dp_2 \wedge dr_2 = \omega$. Let us check also that the translation action preserves β. The translation action associated to an element g of the Lie group $(\mathbb{R}^3, +)$ is

$$S_g : (r_1, r_2, p_1, p_2) \mapsto (r_1 + g, r_2 + g, p_1, p_2),$$

which is the natural consequence of translating the coordinate system in three-space by $\mathbf{g}$. If we pull back β under this symmetry we find

$$
\begin{aligned}
S_{\mathbf{g}}^* \beta &= \mathbf{p}_1 d(\mathbf{r}_1 + \mathbf{g}) + \mathbf{p}_2 d(\mathbf{r}_2 + \mathbf{g}) \\
&= \mathbf{p}_1 \, d\mathbf{r}_1 + \mathbf{p}_1 \, d\mathbf{g} + \mathbf{p}_2 \, d\mathbf{r}_2 + \mathbf{p}_2 \, d\mathbf{g} \\
&= \mathbf{p}_1 \, d\mathbf{r}_1 + \mathbf{p}_2 \, d\mathbf{r}_2 = \beta,
\end{aligned}
$$

since $\mathbf{g}$ is constant.

Exercise 113 *Show that the translation action does not preserve the one-form* $\mathbf{r}_1 d\mathbf{p}_1 + \mathbf{r}_2 d\mathbf{p}_2$.

Another example is the rotation action on the cotangent bundle of $\mathbb{R}^3$. Here we have coordinates $(\mathbf{r}, \mathbf{p})$ and $\beta = \mathbf{p}d\mathbf{r}$. Associated to each element g of the Lie group $SO(3)$ is the symmetry

$$
S_g : (\mathbf{r}, \mathbf{p}) \mapsto (g\mathbf{r}, \mathbf{r}g^T),
$$

the natural consequence of rotating the coordinate system in three-space by g. We find

$$
S_g^* \beta = \mathbf{p}g^T d(g\mathbf{r}) = \mathbf{p}g^T g d\mathbf{r} = \mathbf{p}d\mathbf{r} = \beta.
$$

So we see that in these two examples β is preserved.

The one-form β gives us a recipe for the momentum map. On any cotangent bundle M we can use β to construct a momentum map Φ by the explicit formula

$$
\Phi^\xi := \iota_{\xi_M} \beta
$$

for any ξ in $\mathfrak{g}$. The proof that Φ is in fact a momentum map for the given group action is a consequence of Cartan's magic formula, $\mathcal{L}_X = \iota_X d + d\iota_X$, where $\mathcal{L}_X$ is the Lie derivative with respect to the vector field X. Because β is preserved by the group action it follows that for any Lie algebra element ξ we have $\mathcal{L}_{\xi_M} \beta = 0$. So by Cartan's magic formula, $d\Phi^\xi = d\iota_{\xi_M} \beta = -\iota_{\xi_M} d\beta = -\iota_{\xi_M} \omega$, and Φ is indeed a momentum map.

Let us finish by constructing momentum maps this way in our two examples. For the translation action we know from Section 7.2 that each vector ξ in $\mathbb{R}^3$ corresponds to a vector field $\xi_M = (\frac{\partial}{\partial \mathbf{r}_1} + \frac{\partial}{\partial \mathbf{r}_2})\xi$. So we have

$$
\Phi^\xi = \iota_{\xi_M} \beta = (\mathbf{p}_1 d\mathbf{r}_1 + \mathbf{p}_2 d\mathbf{r}_2)\left(\left(\frac{\partial}{\partial \mathbf{r}_1} + \frac{\partial}{\partial \mathbf{r}_2}\right)\xi\right) = (\mathbf{p}_1 + \mathbf{p}_2)\xi,
$$

and hence $\Phi(\mathbf{r}_1, \mathbf{r}_2, \mathbf{p}_1, \mathbf{p}_2) = \mathbf{p}_1 + \mathbf{p}_2$, as expected. For the rotation action we have $\xi_M = \frac{\partial}{\partial \mathbf{r}} \xi \mathbf{r} - \mathbf{p} \xi \frac{\partial}{\partial \mathbf{p}}$, so

$$\Phi^{\xi} = \iota_{\xi_M} \beta = (\mathbf{p} d\mathbf{r})(\frac{\partial}{\partial \mathbf{r}} \xi \mathbf{r} - \mathbf{p} \xi \frac{\partial}{\partial \mathbf{p}}) = \mathbf{p} \xi \mathbf{r} = \mathbf{p}(\tilde{\xi} \times \mathbf{r}) = \tilde{\xi}^T (\mathbf{r} \times \mathbf{p}^T),$$

which matches Equation 7.2. So in the case of our two most familiar examples, the construction of the momentum map outlined in this section succeeds as we expected.

The construction we have just outlined works for cotangent bundles, but not for arbitrary symplectic manifolds. From the late 1980s up until the present there has been some progress in other cases. For the circle group there are some topological conditions on the manifold and some conditions on the fixed points of the action that guarantee the existence of a momentum map. There are also various examples of circle actions on symplectic manifolds that do not have momentum maps. For a review of these results, see McDuff and Salamon [MS, Section 5.1]. Ginzburg has results for other compact groups [Gi].

Exercise 114 *Let T^2 be the two-dimensional torus, $T^2 = S^1 \times S^1$. Parametrize the first circle by the angle θ and the second by the angle ψ. Let $(\mathbb{R}, +)$ act on T^2 by*

$$S_t : (\theta, \psi) \mapsto (\theta, \psi + t)$$

Describe this action geometrically. Extra credit: *Show that this action has no momentum map.*

8
Reduction and The Two-Body Problem

The goal of this chapter is to analyze the motion of two massive particles (such as the sun and Mars) interacting with one another. We assume that all forces other than mutual gravitational attraction are negligible.

Throughout this chapter we are symplectic geometers. We want to rewrite the two major simplifications (which we call *symplectic reductions*, or *reductions*), in a way that shows how similar they are to one another and to other simplifications in mathematics and physics. Reduction is a powerful and common technique in symplectic geometry and mechanics; we hope that working through one fundamental and historically important example in detail will help us read and understand the modern literature. In the last section we briefly discuss some more modern applications of symplectic reduction.

First we introduce the phase space of two bodies and the Hamiltonian function for gravitational interaction. Consider the symplectic manifold $M_0 = (\mathbb{R}^3 \times \mathbb{R}^3 \times (\mathbb{R}^3)^* \times (\mathbb{R}^3)^*, \omega_0)$, where ω_0 is the canonical symplectic form. That is, we coordinatize M by $(\mathbf{r}_1, \mathbf{r}_2, \mathbf{p}_1, \mathbf{p}_2)$, where $\mathbf{r}_1$ and $\mathbf{r}_2$ are column vectors in $\mathbb{R}^3$ and $\mathbf{p}_1$ and $\mathbf{p}_2$ are row vectors in $(\mathbb{R}^3)^*$. Then, using notation introduced in Section 5.6, we define

$$\omega_0 = d\mathbf{p}_1 \wedge d\mathbf{r}_1 + d\mathbf{p}_2 \wedge d\mathbf{r}_2.$$

Let G, m_1 and m_2 be strictly positive constants. We consider the Hamiltonian function

$$H_0 : \qquad\qquad \mathbb{R}^{12} \to \mathbb{R}$$
$$(\mathbf{r}_1, \mathbf{r}_2, \mathbf{p}_1, \mathbf{p}_2) \mapsto \frac{1}{2}\left(\frac{|\mathbf{p}_1|^2}{m_1} + \frac{|\mathbf{p}_2|^2}{m_2}\right) - \frac{Gm_1 m_2}{|\mathbf{r}_1 - \mathbf{r}_2|}.$$

Notice that by giving the manifold, its symplectic form and this function, we have (implicitly, by the theory of Hamiltonian flows described in Chapter 4) specified a system of ordinary differential equations on the manifold. In Chapter 1 we started with twelve scalar differential equations and rewrote them after each simplification. We will see in the current chapter that the theory of symplectic geometry lets us avoid writing the equations down in coordinates until the very end. (To be fair to modern physicists, we note that they also avoid writing and rewriting the differential equations. They typically achieve this by using the Lagrangian formulation of mechanics, but that is another story. For details, see [Ar89] or [Go, Chapters 2 and 3].)

8.1   First Reduction

The action of the three-dimensional group of translations (translating $\mathbf{r}_1$ and $\mathbf{r}_2$ but fixing $\mathbf{p}_1$ and $\mathbf{p}_2$) preserves both the symplectic form and the Hamiltonian. Because this action has an associated *momentum map*, we can perform symplectic reduction. Because the Hamiltonian function is invariant under the translation action, the symplectic reduction yields a reduced Hamiltonian system.

For any $\mathbf{g} \in \mathbb{R}^3$ we can define a map from the phase space $(\mathbb{R}^{12}, \omega_0)$ to itself:

$$(\mathbf{r}_1, \mathbf{r}_2, \mathbf{p}_1, \mathbf{p}_2) \mapsto (\mathbf{r}_1 + \mathbf{g}, \mathbf{r}_2 + \mathbf{g}, \mathbf{p}_1, \mathbf{p}_2).$$

This is a group action of the Lie group $(\mathbb{R}^3, +)$, as the reader can check from Definition 19 of Section 5.4. This action is symplectic: for any constant $\mathbf{g} \in \mathbb{R}^3$ we have $d\mathbf{g} = 0$ and hence

$$d\mathbf{p}_1 \wedge d(\mathbf{r}_1 + \mathbf{g}) + d\mathbf{p}_2 \wedge d(\mathbf{r}_2 + \mathbf{g}) = d\mathbf{p}_1 \wedge d\mathbf{r}_1 + d\mathbf{p}_2 \wedge d\mathbf{r}_2.$$

For a more formal and detailed justification of this calculation, see Section 5.6. We showed in Section 7.2 that the corresponding momentum map is

$$\Phi_0 : \qquad\qquad \mathbb{R}^{12} \to (\mathbb{R}^3)^*$$
$$(\mathbf{r}_1, \mathbf{r}_2, \mathbf{p}_1, \mathbf{p}_2) \mapsto \mathbf{p}_1 + \mathbf{p}_2.$$

Note that the value of Φ_0 is the total linear momentum of the two particles.

We will perform the reduction in abstract language before seeing what it looks like explicitly for this stage of the two-body problem. Pick any λ in the *momentum space* $(\mathbb{R}^3)^*$. Consider $\Phi_0^{-1}(\lambda)$, i.e., all the points in phase space $(\mathbb{R}^{12})$ with momentum λ. Using the notion of a quotient space defined in Definition 21 of Section 5.5 we set

$$M_1 := \Phi_0^{-1}(\lambda)/(\mathbb{R}^3, +).$$

That is, we identify all points that differ by a translation of the $\mathbf{r}$'s within the set of points with momentum λ. So, for instance, the point in $\mathbb{R}^{12}$ representing the first particle at the origin and the second particle at position $(0, 0, 1)^T$, both moving at speed 1 in the positive x-direction would be identified with any point representing a pair of particles one unit apart, the second directly above the first, both moving with speed 1 in the positive x-direction.

There is a natural *reduced symplectic form* ω_1 on M_1, which we will calculate below explicitly. Its existence is guaranteed by the theory of symplectic reduction. The symplectic manifold (M_1, ω_1) is called the *reduced phase space*. It is often written

$$M_0//_\lambda\mathbb{R}^3$$

and pronounced "M_0 mod mod $\mathbb{R}^3$ (at λ)." To get the *reduced Hamiltonian H_1* we write the formula for H_0, namely,

$$H_0 = \frac{1}{2}\left(\frac{|\mathbf{p}_1|^2}{m_1} + \frac{|\mathbf{p}_2|^2}{m_2}\right) - \frac{Gm_1m_2}{|\mathbf{r}_1 - \mathbf{r}_2|},$$

and note that its value is invariant under simultaneous translation of the $\mathbf{r}$'s. So the formula defines a function $H_1 : M_1 \to \mathbb{R}$.

Now let us work this reduction out explicitly. Pick $\lambda \in \mathbb{R}^3$. Considering $\Phi_0^{-1}(\lambda)$ means that we restrict our attention to situations where $\mathbf{p}_1 + \mathbf{p}_2 = \lambda$. That is,

$$\Phi_0^{-1}(\lambda) = \{(\mathbf{r}_1, \mathbf{r}_2, \mathbf{p}_1, \mathbf{p}_2) \in \mathbb{R}^{12} : \mathbf{p}_1 + \mathbf{p}_2 = \lambda\}.$$

Next we will find coordinates on $M_0//_\lambda\mathbb{R}^3$. Consider the function

$$q : \Phi_0^{-1}(\lambda) \to \mathbb{R}^6$$
$$(\mathbf{r}_1, \mathbf{r}_2, \mathbf{p}_1, \mathbf{p}_2) \mapsto (\mathbf{r}_1 - \mathbf{r}_2, \tfrac{1}{m_1+m_2}(m_2\mathbf{p}_1 - m_1\mathbf{p}_2)). \tag{8.1}$$

We can think of this function as the quotient map described in Section 5.5. First note that for any $\mathbf{g} \in \mathbb{R}^3$ we have $q(\mathbf{r}_1 + \mathbf{g}, \mathbf{r}_2 + \mathbf{g}, \mathbf{p}_1, \mathbf{p}_2) = q(\mathbf{r}_1, \mathbf{r}_2, \mathbf{p}_1, \mathbf{p}_2)$ for each element of $\mathbb{R}^{12}$. So any two elements of one orbit of $(\mathbb{R}^3, +)$ evaluate to the same value of q. On the other hand, we can reconstruct $\mathbf{r}_1$ and $\mathbf{r}_2$ (up to a translation) and $\mathbf{p}_1$ and $\mathbf{p}_2$ from $\mathbf{r}$ and $\mathbf{p}$ (and, of course, the fixed row vector λ):

$$\mathbf{p}_1 = \frac{m_1\lambda}{m_1 + m_2} + \mathbf{p}$$

$$\mathbf{p}_2 = \frac{m_2\lambda}{m_1 + m_2} - \mathbf{p},$$

$$(\mathbf{r}_1, \mathbf{r}_2) \sim \frac{1}{2}(\mathbf{r}, -\mathbf{r}),$$

since $\mathbf{r}_1 = \frac{1}{2}\mathbf{r} + \frac{1}{2}(\mathbf{r}_1 + \mathbf{r}_2)$ and $\mathbf{r}_2 = -\frac{1}{2}\mathbf{r} + \frac{1}{2}(\mathbf{r}_1 + \mathbf{r}_2)$. A more organized way to express these relationships is to write that

$$(\mathbf{r}_1, \mathbf{r}_2, \mathbf{p}_1, \mathbf{p}_2) \text{ is equivalent to } \left(\frac{1}{2}\mathbf{r}, -\frac{1}{2}\mathbf{r}, \frac{m_1\lambda}{m_1 + m_2} + \mathbf{p}, \frac{m_2\lambda}{m_1 + m_2} - \mathbf{p}\right).$$
(8.2)

Note that this last expression depends only on constant quantities and our new coordinates. We conclude that $M_1 \cong \mathbb{R}^6$. If we write $(\mathbf{r}, \mathbf{p}) := q(\mathbf{r}_1 + \mathbf{g}, \mathbf{r}_2 + \mathbf{g}, \mathbf{p}_1, \mathbf{p}_2)$ then $(\mathbf{r}, \mathbf{p})$ can serve as coordinates on the quotient space M_1.

Let us calculate the natural symplectic form ω_1 on M_1. Using Equations 8.1 and 8.2 we have

$$d\mathbf{p}_1 \wedge d\mathbf{r}_1 + d\mathbf{p}_2 \wedge d\mathbf{r}_2$$
$$\overset{?!}{=} d\left(\tfrac{m_1\lambda}{m_1+m_2} + \mathbf{p}\right) \wedge \tfrac{1}{2}d\mathbf{r} + d\left(\tfrac{m_2\lambda}{m_1m_2} - \mathbf{p}\right) \wedge \left(-\tfrac{1}{2}d\mathbf{r}\right)$$
(8.3)
$$= d\mathbf{p} \wedge d\mathbf{r}.$$

So it is consistent to define $\omega_1 := d\mathbf{p} \wedge d\mathbf{r}$.

Note that the first "equality" in the above equation is fishy. All we have is an equivalence relation, not an equality. For instance, it is not necessarily true that $\mathbf{r}_1 = \frac{1}{2}\mathbf{r}$. We can state the precise relationship between ω_0 and ω_1 by using the quotient map $q : \Phi_0^{-1}(\lambda) \to M_1$ (see Section 5.5) and the inclusion map $\Gamma : \Phi_0^{-1}(\lambda) \to M_0$ defined by $\Gamma(m) := m$. We have

$$q^*\omega_1 = q^*(d\mathbf{r} \wedge d\mathbf{p}) = d(\mathbf{r}_1 - \mathbf{r}_2) \wedge d(\frac{m_2\mathbf{p}_1 - m_1\mathbf{p}_2}{m_1 + m_2}) = d(\mathbf{r}_1 - \mathbf{r}_2) \wedge d\mathbf{p}_1,$$

where the last equality depends on the fact that $\mathbf{p}_1 + \mathbf{p}_2 = \lambda$ on $\Phi_0^{-1}(\lambda)$ and hence $d\mathbf{p}_2 = -d\mathbf{p}_1$ there. Similarly, we have

$$\Gamma^*\omega_0 = \Gamma^*(d\mathbf{r}_1 \wedge d\mathbf{p}_1 + d\mathbf{r}_2 \wedge d\mathbf{r}_2) = (d\mathbf{r}_1 - d\mathbf{r}_2) \wedge d\mathbf{p}_1.$$

So $q^*\omega_1 = \Gamma^*\omega_0$. The Marsden–Weinstein–Meyer theorem (Theorem 3 of Section 8.5) of symplectic reduction says that such a symplectic form always exists on the quotient space and it is unique. In other words, if we had written any other of the (infinite variety of) formulas for elements of the equivalence class of $(\mathbf{r}_1, \mathbf{r}_2, \mathbf{p}_1, \mathbf{p}_2)$ in terms of $(\mathbf{r}, \mathbf{p})$ we would have calculated the same symplectic form in Equation 8.3.

Finally, we calculate the reduced Hamiltonian, which we call H_1. We require that $H_0 = H_1 \circ \Gamma$. Since the Hamiltonian function

$$H_0 : \qquad\qquad \mathbb{R}^6 \to \mathbb{R}$$

$$(\mathbf{r}_1, \mathbf{r}_2, \mathbf{p}_1, \mathbf{p}_2) \mapsto \tfrac{1}{2}\left(\tfrac{|\mathbf{p}|^2}{m_1} + \tfrac{|\mathbf{p}_2|^2}{m_2} \right) - \tfrac{G m_1 m_2}{|\mathbf{r}_1 - \mathbf{r}_2|},$$

is preserved by the group action we can take

$$H_1 : \qquad\qquad \mathbb{R}^6 \to \mathbb{R}$$

$$(\mathbf{r}, \mathbf{p}) \mapsto \tfrac{1}{2}\left(\tfrac{|\mathbf{p}|^2}{\mu} + \tfrac{|\lambda|^2}{M} \right) - \tfrac{G M \mu}{|\mathbf{r}|},$$

where $\mu := m_1 m_2 / (m_1 + m_2)$ is the reduced mass and $M := m_1 + m_2$ is the total mass of the system.

We have completed the first reduction. We've constructed a manifold M_1 of lower dimension with a symplectic form ω_1 and a Hamiltonian function H_1. Together these define a Hamiltonian flow on M_1, by the general procedure discussed in Section 4.1. Note that the momentum map Φ_0 and the invariance of H_0 under the group action were essential to this construction.

The reduction we have just completed corresponds to the physicists' passage to center-of-mass coordinates, which was introduced by Newton (see Section 1.1). Thus the modern technique of symplectic reduction is as old as the *Principia*! We pause to reflect in more detail on the relationship between passage to center-of-mass coordinates and symplectic reduction.

In the physics style derivation of Chapter 1 the use of center-of-mass coordinates gave us two equations: a momentum equation $\mathbf{p}_1 + \mathbf{p}_2 = 0$ and a position equation $m_1 \mathbf{r}_1 + m_2 \mathbf{r}_2 = 0$. The momentum equation corresponds to the first step of the reduction, the consideration of $\Phi_0^{-1}(\lambda)$ for a fixed value λ of the total momentum. The correspondence is exact if $\lambda = 0$; the discrepancy for nonzero values of λ can be thought of as a symptom of a philosophical difference between mathematics and physics: mathematicians are willing to keep track of a relatively unimportant parameter in deference to a general method, while physicists are willing to sacrifice generalizability in order to keep the solution of the problem at hand free of unnecessary clutter. In other examples of symplectic reduction (such as our second reduction, below, using the angular momentum) the structure of the reduced space does depend on the value of the momentum map; mathematicians tend to want to emphasize the relationship to other instances of the general theory, while physicists tend to want to emphasize the special features of the physical problem at hand.

The position equation corresponds to the second part of the reduction, the identification of states differing by a translation. While reducing we considered the equivalence classes of all points in $\mathbb{R}^{12}$ differing by simultaneous translations of $\mathbf{r}_1$ and $\mathbf{r}_2$. In each such equivalence class there is exactly one point satisfying $m_1\mathbf{r}_1 + m_2\mathbf{r}_2 = 0$ (obtained by taking $\mathbf{g} = -(m_1\mathbf{r}_1 + m_2\mathbf{r}_2/(m_1 + m_2))$). So we can label equivalence classes by pairs $(\mathbf{r}_1, \mathbf{r}_2)$ satisfying the position equation. We can describe this situation more formally: the set of points satisfying the position equation is a *cross-section* of the equivalence classes considered in the second reduction.

Because the symplectic form ω_1 is the usual form on $\mathbb{R}^3 \times (\mathbb{R}^3)^*$ (see Section 2.5) and our Hamiltonian has a kinetic energy term $|\mathbf{p}|^2/(2\mu)$, we are free to interpret this reduced system as describing the motion of one particle of mass μ in three-space, moving under the influence of a potential energy

$$\frac{1}{2}\frac{|\lambda|^2}{M} - \frac{GM\mu}{|\mathbf{r}|}.$$

This is an example of what physicists call a *particle in a central force field*. This particle's position is $\mathbf{r}$. Its momentum is $\mathbf{p}$.

In fancier terminology: we used translational symmetry to reduce the Hamiltonian system (M_0, ω_0, H_0) at λ to (M_1, ω_1, H_1). Note that M_1, ω_1 and H_1 all depend on λ. We often call (M_0, ω_0, H_0) the *upstairs system* and (M_1, ω_1, H_1) the *downstairs system* or the *reduced system*. It is a theorem of symplectic geometry and mechanics that the flow generated in M_1 by H_1 downstairs corresponds to the flow generated in M_0 by H_0. See Theorem 4 in Section 8.5. In other words, if we can solve the (usually simpler) downstairs system, then we gain a lot of information about the upstairs system. In many cases, solving the downstairs system leads to solving the upstairs system outright. See [AM, Theorem 4.3.5 and ff.].

In summary, our reduced Hamiltonian system has phase space $M_1 = \mathbb{R}^3 \times (\mathbb{R}^3)^* = \{(\mathbf{r}, \mathbf{p})\}$ with symplectic form $\omega_1 = dp \wedge dr$ and Hamiltonian $H_1(\mathbf{r}, \mathbf{p}) = \frac{1}{2}\left(\frac{|\mathbf{p}|^2}{\mu} + \frac{|\lambda|^2}{M}\right) - \frac{GM\mu}{|\mathbf{r}|}$, where $\lambda \in \mathbb{R}^3$, $\mu \in \mathbb{R}$ and $M \in \mathbb{R}$ are constants representing the total momentum, the reduced mass and the total mass, respectively.

8.2   Second Reduction

The rotation action of the group $SO(3)$ of rotations of three-space acts on M_1, as we have seen in Section 5.6. This action preserves the symplectic form and the Hamiltonian function, and it has a momentum map, as we have seen in Section 7.2. Recall that a rotation matrix $g \in SO(3)$ acts by

$$S_g : (\mathbf{p}, \mathbf{r}) \mapsto (\mathbf{p}g^T, g\mathbf{r})$$

and the momentum map for this action is

$$
\begin{aligned}
\Phi_1 : \quad & M_1 \to so(3) \\
& (\mathbf{p}, \mathbf{r}) \mapsto \left((\mathbf{r}\mathbf{p})^T - \mathbf{r}\mathbf{p} \right).
\end{aligned}
$$

where we have used the identification of the Lie algebra $so(3)$ with its dual $so(3)^*$, as in Formula 7.1 in Section 7.1. In coordinates,

$$
(\mathbf{r}\mathbf{p})^T - \mathbf{r}\mathbf{p} = \begin{pmatrix} 0 & -(r_x p_y - r_y p_x) & (r_z p_x - r_x p_z) \\ (r_x p_y - r_y p_x) & 0 & -(r_y p_z - r_z p_y) \\ -(r_z p_x - r_x p_z) & (r_y p_z - r_z p_y) & 0 \end{pmatrix}.
$$

Note that the entries in this matrix are the entries in the angular momentum vector $\mathbf{r} \times \mathbf{p}^T$.

Enough preliminaries. Let's reduce! Fix any skew symmetric matrix L. Let $\tilde{L}$ be the corresponding three-vector (see Definition 25 of chapter 6). Consider $\Phi_1^{-1}(L)$, i.e., the set of all points in the phase space $M_1 = \mathbb{R}^6$ with angular momentum $\tilde{L}$. Define

$$M_2 := \Phi_1^{-1}(L)/SO(3).$$

That is, we identify points in $\Phi_1^{-1}(L)$ that differ by an overall rotation.

Notice that some rotations will change the angular momentum vector and will carry $(\mathbf{r}, \mathbf{p})$ to a pair with different angular momentum. We will ignore these; we are building equivalence classes inside $\Phi_1^{-1}(L)$, not on all of M_1. For instance, the situation with the particle on the positive x-axis one unit from the center with momentum vector three units long in the positive y-direction would be equivalent to the situation with the particle one unit from the center on the positive y-axis with momentum vector three units long pointing in the negative x-direction. See Figure 8.1.

At this point we have to split the analysis into two cases, depending on the value of the angular momentum. In the first reduction (Section 8.1) we were lucky enough to be able to treat all the reduced spaces uniformly. But

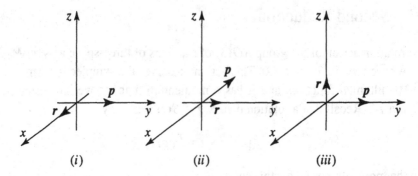

Figure 8.1. (i) and (ii) are equivalent; (iii) has a different angular momentum

there is nothing to guarantee in general that different values of the momentum map should yield the same reduced space, as the angular momentum example illustrates. We will see that the structure of the reduced space M_2 when $L = 0$ is different from the structure when $L \neq 0$. In particular, when $L = 0$ the reduced space is not a manifold. We will consider the zero angular momentum case in Section 8.4. In the rest of the current section we analyze the case $L \neq 0$.

Let us find coordinates on M_2 so that we can write the reduced symplectic form explicitly. Note that $\Phi_1^{-1}(L)$ is three-dimensional. It is cut out of $\mathbb{R}^6$ by the three independent scalar equations $\mathbf{r} \times \mathbf{p}^T = \tilde{L}$.

Exercise 115 *Use the implicit function theorem to show that for any nonzero* $\tilde{L} \in \mathbb{R}^3$ *the set*

$$S_{\tilde{L}}\{(\mathbf{r}, \mathbf{p}) \in \mathbb{R}^6 : \mathbf{r} \times \mathbf{p}^T = \tilde{L}\}$$

can be parametrized locally differentiably near any point by three parameters. In other words, the set is a three-dimensional manifold. Can you find a global parametrization? What can you say about the set S_0? *Is it a manifold?*

Here are coordinates on $M_2 := \Phi_1^{-1}(L)/SO(3)$:

$$\rho := |\mathbf{r}|$$
$$\sigma := \frac{\mathbf{pr}}{|\mathbf{r}|}$$

It might have been more natural to use π instead of σ, since π is the Greek letter corresponding to $\mathbf{p}$, but there is a limit to the service we can require of one poor symbol. We'll see that the coordinate σ ends up playing the role of a momentum. Notice that σ is the signed length of the component of the momentum $\mathbf{p}$ in the radial direction. As in the previous reduction, we can think of these formulas as defining the quotient map $q : (\mathbf{r}, \mathbf{p}) \mapsto (\rho, \sigma)$.

Next we show that (up to a rotation) we can recover $(\mathbf{r}, \mathbf{p})$ from (ρ, σ). Specifically, we show that

$$\left(\begin{pmatrix} \rho \\ 0 \\ 0 \end{pmatrix}, \begin{pmatrix} \sigma & 0 & \frac{-|\tilde{L}|}{\rho} \end{pmatrix} \right) \text{ is equivalent to } (\mathbf{r}, \mathbf{p}) \qquad (8.4)$$

under the rotation action of $SO(3)$ introduced in Section 5.6. Note that the nonzero vectors $\mathbf{r}$, $\tilde{L}$ and $\mathbf{r} \times \tilde{L}$ are mutually orthogonal and obey the right-hand rule. Hence the matrix

$$g := \begin{pmatrix} \dfrac{\mathbf{r}}{|\mathbf{r}|} & \dfrac{\tilde{L}}{|\tilde{L}|} & \dfrac{\mathbf{r} \times \tilde{L}}{|\mathbf{r} \times \tilde{L}|} \end{pmatrix}$$

is an element of $SO(3)$. But by the definition of ρ we have

$$g \begin{pmatrix} \rho \\ 0 \\ 0 \end{pmatrix} = \frac{\mathbf{r}}{|\mathbf{r}|} \rho = \mathbf{r},$$

while, by the definitions of σ, ρ and $\tilde{L}$ and the geometric meaning of the cross-product,

$$\mathbf{p}g = \begin{pmatrix} \dfrac{\mathbf{pr}}{|\mathbf{r}|} & \dfrac{\mathbf{p}\tilde{L}}{|\tilde{L}|} & \dfrac{\mathbf{p}(\mathbf{r} \times \tilde{L})}{|\mathbf{r} \times \tilde{L}|} \end{pmatrix} = \begin{pmatrix} \sigma & 0 & \dfrac{-|\tilde{L}|}{\rho} \end{pmatrix}.$$

Since $g \in SO(3)$ we have $g^{-1} = g^T$, and hence $(\sigma, 0, -|\tilde{L}|/\rho)g^T = \mathbf{p}$. So g effects the desired equivalence.

Exercise 116 Check that $\frac{\mathbf{pr}}{|\mathbf{r}|} = \sigma$, that $\mathbf{p}\tilde{L} = 0$ and that $\frac{\mathbf{p}(\mathbf{r}\times\tilde{L})}{|\mathbf{r}\times\tilde{L}|} = \frac{-|\tilde{L}|}{\rho}$.

The point of the calculation we have just finished is that (ρ, σ) are good coordinates on M_2. Note, however, that only strictly positive values of ρ arise, while for any fixed $\rho > 0$ any $\sigma \in \mathbb{R}$ can arise. The reader should verify this, recalling that $\tilde{L} \neq 0$. So $M_2 \cong \mathbb{R}^+ \times \mathbb{R}$. In other words, the reduced space M_2 is the open right half plane. See Figure 8.2.

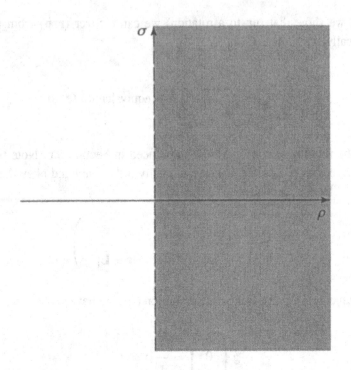

Figure 8.2. The manifold M_2, the open right half plane

Let us calculate the reduced symplectic form on M_2. Using the equivalence (8.4) we calculate (as we did in the first reduction)

$$dp \wedge dr \stackrel{?!}{=} d \left(\sigma \quad |\tilde{L}|/\rho \quad 0 \right) \wedge d \begin{pmatrix} \rho \\ 0 \\ 0 \end{pmatrix}$$

$$= d\sigma \wedge d\rho.$$

So $\omega_2 = d\sigma \wedge d\rho$. In other words, the symplectic form on M_2 is the usual area form. More formally, using the quotient map q and the inclusion map Γ of $\Phi_2^{-1}(L)$ into M_1, a careful calculation using the fact that $d(\mathbf{r} \times \mathbf{p}^T) = 0$ on $\Phi_2^{-1}(L)$ yields $\Gamma^*(d\mathbf{p} \wedge d\mathbf{r}) = q^*(d\sigma \wedge d\rho)$.

Exercise 117 *Verify that* $\Gamma^*(d\mathbf{p} \wedge d\mathbf{r}) = q^*(d\sigma \wedge d\rho)$ *by direct calculation.*

The reduced Hamiltonian is

$$H_2: \qquad \mathbb{R}^+ \times \mathbb{R} \to \mathbb{R}$$

$$(\rho, \sigma) \mapsto \tfrac{\lambda^2}{2M} + \tfrac{\sigma^2}{2\mu} + \tfrac{|\tilde{L}|^2}{2\mu\rho^2} - \tfrac{GM\mu}{\rho}.$$

We have completed the second reduction. This reduction corresponds to the second simplification of Section 1.2, the exploitation of the conservation of angular momentum.

We can interpret the twice-reduced system physically as a particle on $\mathbb{R}^+$ subject to two opposing forces. If we subtract the kinetic energy and constant terms from the Hamiltonian we are left with potential energy

$$\frac{\left|\tilde{\mathbf{L}}\right|^2}{2\mu\rho^2} - \frac{GM\mu}{\rho}.$$

The first term corresponds to a force of strength proportional to ρ^{-3} repelling the particle from $\rho = 0$, while the second term corresponds to a force of strength proportional to ρ^{-2} attracting the particle toward $\rho = 0$. For large ρ, the attractive force is more significant; for small ρ the repellent force dominates. We have

$$\frac{\left|\tilde{\mathbf{L}}\right|^2}{2\mu} \leq H_2\rho^2 + GM\mu\rho$$

and so, because the Hamiltonian H_2 is constant along any trajectory and because $\left|\tilde{\mathbf{L}}\right| \neq 0$, the position ρ is bounded away from 0 along any trajectory. This means that the position of a particle might oscillate between two strictly positive values, or it might increase without bound, but it will never "crash" into the forbidden $\rho = 0$. We can conclude that our two original point masses will never crash into one another if they have nonzero angular momentum about the center of mass.

In summary, our twice-reduced Hamiltonian system has phase space $M_2 = \mathbb{R}^+ \times \mathbb{R} = \{(\rho, \sigma) : \rho > 0\}$ with symplectic form $\omega_2 = d\sigma \wedge d\rho$ and Hamiltonian $H_2(\rho, \sigma) = \frac{\lambda^2}{2M} + \frac{\sigma^2}{2\mu} + \frac{\left|\tilde{\mathbf{L}}\right|^2}{2\mu\rho^2} - \frac{GM\mu}{\rho}$. Note that $\lambda \in \mathbb{R}^3, \tilde{\mathbf{L}} \in \mathbb{R}^3, M \in \mathbb{R}^+$ and $\mu \in \mathbb{R}^+$ are all constant.

8.3   Recovering Kepler's Laws

Our recovery of Kepler's laws in Section 1.3 (which the reader may now wish to review) depended only on the conservation of angular momentum, conservation of linear momentum and Equation 1.7, which we recall for the reader's convenience:

$$\frac{d^2\rho}{dt^2} - \rho\left(\frac{d\theta}{dt}\right)^2 + \frac{GM}{\rho^2} = 0.$$

In the current chapter we have already established conservation of linear and angular momentum, so it remains only to derive Equation 1.7.

By the theory of symplectic reduction, we can use the Hamiltonian H_2 and the symplectic form ω_2 to write down the differential equations for the time flow of ρ and σ, as described in Section 4.1.

Exercise 118 *Show that the Hamiltonian flow is given by*

$$\frac{d\sigma}{dt} = \frac{|\tilde{\mathbf{L}}|^2}{\mu\rho^3} - \frac{GM\mu}{\rho^2}$$

$$\frac{d\rho}{dt} = \frac{\sigma}{\mu}.$$

Combining the two Hamiltonian flow equations from the exercise we find

$$\frac{d^2\rho}{dt^2} - \frac{|\tilde{\mathbf{L}}|^2}{\mu^2\rho^3} + \frac{GM}{\rho^2} = 0.$$

So to derive Equation 1.7 it suffices to show that $\mu\rho^2\frac{d\theta}{dt} = |\tilde{\mathbf{L}}|$. But, as noted in Section 1.3 this last equality is just the conservation of angular momentum written in terms of $\frac{d\theta}{dt}$. From this point on, the derivation of Kepler's Laws proceeds exactly as in Section 1.3.

Exercise 119 *Show that the* Runge-Lenz vector

$$\mathbf{R} := \mathbf{p}^T \times \tilde{\mathbf{L}} - GM\mu^2\frac{\mathbf{r}}{|\mathbf{r}|}$$

is constant under the flow. Show that this vector is parallel to the semimajor axis of the ellipse. For more information on the Runge Lenz vector, see the article by Iglesias [I] or the book by Guillemin and Sternberg [GS90].

8.4 Zero Angular Momentum

When the angular momentum is zero, all rotations preserve angular momentum. This case corresponds to a situation where the momenta of the two original particles are parallel to the line joining them (in center-of-mass coordinates). The motion takes place along a line through the origin; the identification of situations via the action of $SO(3)$ is the mathematical way of saying that it does not much matter which line the motion is on. The only curious situation is the point $(\mathbf{0}, \mathbf{0})$ in M_1, which represents a point at the origin with no linear momentum at all. Such a point is "moving" on all lines through the origin simultaneously.

Exercise 120 *Find the orbit of $SO(3)$ through $(\mathbf{0}, \mathbf{0})$. Find the orbit of $SO(3)$ through an arbitrary nonzero point in $\Phi_1^{-1}(0)$.*

Are we making a mountain out of a molehill? As mathematicians we must be careful. In fact, the point $(\mathbf{0}, \mathbf{0})$ is a troublemaker: the quotient space $M_1//_0SO(3) := \Phi_1^{-1}(0)/SO(3)$ is not a manifold because there are no good coordinates near the point in $M_1//_0SO(3)$ representing the orbit through $(\mathbf{0}, \mathbf{0})$. Loosely stated, the problem is that, up to $SO(3)$ symmetry, the only direction to travel in phase space starting at $(\mathbf{0}, \mathbf{0})$ is "out." In other words, we can use up the $SO(3)$ symmetry by thinking of every particle as being on the nonnegative x-axis, but the origin on the x-axis does not have a neighborhood that looks like Euclidean space. Readers familiar with differential geometry are invited to turn this intuition into a proof in Exercise 121.

Exercise 121 (Optional) *Show that $\Phi_1^{-1}(0)$ is not a manifold.*
Hint: Show that each coordinate direction in $\mathbb{R}^6$ is tangent to $\Phi_1^{-1}(0)$ at $(\mathbf{0}, \mathbf{0})$ but $\Phi_1^{-1}(0)$ cannot be six-dimensional. *Also show that $M_1//_0SO(3)$ is not a manifold.*

There are mathematical techniques for analyzing spaces such as $M_1//_0SO(3)$ that are almost but not quite manifolds. One particular class of spaces, called *orbifolds*, has been studied in detail. An orbifold is very much like a manifold in that it must have a "nice" coordinate patch in a neighborhood of each point, but the exact definition of "nice" is less restrictive. For a manifold, nice coordinate systems are diffeomorphisms from an open subset of $\mathbb{R}^n$; for an orbifold, nice coordinate systems are diffeomorphisms from an open set in a quotient space $\mathbb{R}^n/G$, where G is a finite group. For an introduction to group actions and momentum maps on symplectic orbifolds (especially if the group is a torus, i.e., a direct product of a finite number of circle groups), see the article by Lerman and Tolman [LT].

We will modify the problem to avoid the complication of introducing orbifolds. We use the collision-free manifold $\hat{M}_0 := \{(\mathbf{r}_1, \mathbf{r}_2, \mathbf{p}_1, \mathbf{p}_2) : \mathbf{r}_1 \neq \mathbf{r}_2\}$, introduced in Section 2.5. With this modification the first reduction yields $\hat{M}_1 = \{(\mathbf{r}, \mathbf{p}) : \mathbf{r} \neq \mathbf{0}\}$. In this case $\hat{M}_2 = \hat{M}_1//_0SO(3)$ is a manifold. In fact, it is the manifold $\mathbb{R}^+ \times \mathbb{R} = \{(\rho, \sigma) : \rho > 0\}$. As in Section 8.2 we have $\omega_2 = d\sigma \wedge d\rho$ and, since $L = 0$, the Hamiltonian function is

$$H_2 = \frac{\lambda^2}{2M} + \frac{\sigma^2}{2\mu} - \frac{GM\mu}{\rho}$$

As the reader has perhaps verified in Exercise 69 in Section 4.4, the differential equation of motion is

$$\frac{d}{dt}\rho = \pm\sqrt{\frac{2}{\mu}(H_2 - \frac{\lambda^2}{2M} + \frac{GM\mu}{\rho})}, \tag{8.5}$$

and its solution (expressing t as a function of the position ρ and the initial time t_0) can be obtained from an integral table or a good symbolic integration program. If $h := H_2 - \frac{\lambda^2}{2M} \geq 0$, i.e., if the total energy H_2 is larger than just the kinetic energy due to the motion of the center of mass, we have (setting $k = GM\mu$)

$$t - t_0 = \pm\frac{1}{\mu k^2\sqrt{2}}\left(\frac{1}{2h^{3/2}}\ln\left|\frac{\sqrt{h + (k/\rho)} - \sqrt{h}}{\sqrt{h + (k/\rho)} + \sqrt{h}}\right| + \frac{1}{kh}\rho\sqrt{h + (k/\rho)}\right).$$

If, on the other hand, the total energy H_2 is smaller than just the kinetic energy due to the motion of the center of mass, we find that

$$t - t_0 = \pm\frac{1}{\mu k^2\sqrt{2}}\left(\frac{1}{kh}\rho\sqrt{h + (k/\rho)} - \frac{1}{(-h)^{3/2}}\arctan\frac{\sqrt{h + (k/\rho)}}{\sqrt{-h}}\right).$$

Exercise 122 *Check that these equations satisfy the differential equation, Equation 8.5. Interpret these equations physically, i.e., describe all possible motions of the planet and match them up with the equations. Remember that $\rho \neq 0$ in this phase space.*

With the explicit solutions we have just obtained, we complete our analysis of the two-body problem. We have formulas that predict motion for every possible physical situation. However, there are natural mathematical questions that go beyond physical considerations. For instance, mathematicians prefer *closed* manifolds, i.e., manifolds we cannot leave by taking limits. Problematic points such as those with $r_1 \neq r_2$ are called *singularities*. It is often enlightening to *regularize the singularities* by finding a diffeomorphism of the manifold in question into and almost onto a closed manifold of the same dimension. Jürgen Moser achieved this for the two-body problem in 1970 [Mo]. He showed that the singularities can be regularized by mapping the manifold M_1 into the unit tangent bundle of the three-dimensional sphere S^3. In addition, he showed that the orbits of the two-body problem flow on M_1 map to the orbits of the geodesic flow on S^3.

8.5   Symplectic Geometry

In this book we have illustrated several theorems of symplectic geometry whose applicability goes far beyond celestial mechanics. In Section 4.4 we saw examples of the preservation of the Hamiltonian function and the symplectic form by the Hamiltonian flow (Theorems 1 and 2 of Section 4.4). We mentioned Darboux's Theorem in Section 2.5. The current chapter illustrates two more theorems. The first published versions of these theorems were in an article by Kenneth Meyer in 1973 [Me]. In 1974 Jerrold Marsden and Alan Weinstein published a stronger version [MW].

The first theorem is known as the Marsden-Weinstein or Marsden-Weinstein-Meyer reduction theorem. To state the technical hypotheses properly we must use some concepts we have not introduced in this text. Readers who wish to understand these hypotheses will need to consult other sources (such as Abraham and Marsden [AM]); other readers may wish to trust that the hypotheses are often satisfied in practice. One essential ingredient is the *coadjoint action* Ad^* of a Lie group G on the dual $\mathfrak{g}^*$ of its Lie algebra.

Theorem 3 *Let* (M, ω) *be a symplectic manifold. Let* G *be a Lie group acting on* (M, ω) *and let* Φ *be a momentum map for the action of* G. *Suppose that the momentum map* Φ *is* Ad^*-*equivariant, i.e., that for every* g *in* G *and for every* ξ *in the Lie algebra* $\mathfrak{g}$ *of* G *we have*

$$\Phi^\xi \circ S_g = \Phi^{g\xi g^{-1}}.$$

Suppose also that λ *is a* regular *value for the momentum map* Φ, *i.e., that the rank of the Jacobian matrix of* Φ *does not drop from its maximum value at any point on* $\Phi^{-1}(\lambda)$. *Suppose further that the isotropy group* $G_\lambda := \{g \in G : Ad^*_g(\lambda) = \lambda\}$ *acts* freely *and* properly *on* $\Phi^{-1}(\lambda)$. *Then* $M_1 := \Phi^{-1}(\lambda)/G_\lambda$ *is a manifold. If we let* q *denote the quotient map from* $\Phi^{-1}(\lambda)$ *to* M_1 *and we let* Γ *denote the inclusion of* $\Phi^{-1}(\lambda)$ *into* M, *there is a unique symplectic form* ω_1 *on* M_1 *such that* $q^*\omega_1 = \Gamma^*\omega$.

In short, the conclusion of the theorem says that the reduced space is a symplectic manifold. For a proof, see [Br, Lecture 7, Theorem 2] or [AM, Theorem 4.3.1].

The astute reader may have noticed that the reduced space in the zero angular momentum case for the two-body problem (Section 8.4) does not satisfy the conclusion of Theorem 3. This is not a counterexample to the theorem because the hypotheses are not satisfied. More precisely, zero is not a regular value for the angular momentum map.

The next theorem relates the Hamiltonian flow upstairs on the original manifold with the Hamiltonian flow downstairs on the reduced space. We have implicitly used the fact that if the group action respects the Hamiltonian function then the reduced Hamiltonian system corresponds to the original Hamiltonian system. Precisely:

Theorem 4 *Suppose the hypotheses of Theorem 3 hold true. Suppose in addition that a function $H : M \to \mathbb{R}$ is preserved under the action of the group G, i.e., $H \circ S_g = H$ for all $g \in G$. Then there is a function $H^\lambda : M_1 \to \mathbb{R}$ such that $H^\lambda \circ q = H$. Furthermore, if Γ_t denotes the Hamiltonian flow of H on (M, ω) and Γ_t^λ denotes the Hamiltonian flow of H^λ on M_1 we have $q \circ \Gamma_t(m) = \Gamma_t^\lambda \circ q(m)$ for every $m \in \Phi^{-1}(\lambda)$.*

8.6   Concluding Remarks

The main message of this book is that the passage to center-of-mass and the exploitation of the conservation of angular momentum in the two-body problem are both examples of symplectic reduction. But while the two-body problem is historically and etymologically important for modern symplectic geometry, it can hardly serve as a motivation for the development of the mathematical constructs introduced in this text. After all, Newton understood and explained this physics problem fully in the 1600s, long before anyone thought of the definitions of manifolds, Lie groups or moment maps. To make the case that symplectic reduction is not just mathematically appealing but actually powerful, one must present examples in which symplectic geometry does more than organize an existing calculation.

Symplectic reduction can be used to simplify the analysis of mechanical systems. For example, consider the *free rigid body*, otherwise known as the *Euler top*. You can easily study this system experimentally: just toss a tennis racket, or a book sealed shut, into the air and consider its motion around its center of mass. The fact that every possible trajectory of the free rigid body is either periodic or *quasi-periodic* (i.e., returns arbitrarily close to its initial state) is made easy by symplectic reduction.

Another application of symplectic reduction is to the study of relative equilibria of mechanical systems. While equilibrium points are points in phase space that are fixed by the Hamiltonian flow, *relative equilibria* are points in the reduced phase space which are fixed by the Hamiltonian flow of the reduced Hamiltonian. These correspond to particularly simple, symmetric trajectories of the original system. For instance, in a system such as the asymmetric rigid body in a constant gravitational field, a relative equilibrium is either a fixed point or a pure rotation of the body around the vertical axis. Just as one can use equilibrium points of simple systems to define separatrices and study global behavior, one can use relative equilibria to study global behavior of more complicated systems. Topological analysis on the reduced space can prove existence of relative equilibria. See, for example, Arnold and Givent', [AG, Section 3.6].

Symplectic reduction can expose links between mechanics and other subjects, such as topology. For instance, if we consider a single massive particle on the plane moving under a force with a quadratic potential function, the energy is a momentum map for the flow, which is a circle action. Performing symplectic reduction at a nonzero value of the momentum map yields the Hopf fibration of the three-sphere, which is well known to topologists. Note that this symplectic reduction does not correspond to any physically simple choice of coordinates on the configuration space.

Symplectic reduction can also expose links between different mechanical systems. For example, one can carry out reduction of the phase space of the free rigid body to obtain the phase space and equations of motion for the planar pendulum. For details and references, see Marsden and Ratiu [MR, Section 15.11].

The momentum map has had an impact on modern mathematics. One major development was the convexity theorem for Hamiltonian torus actions. This theorem says that if (M, ω) is a compact, connected symplectic manifold, then the image of the momentum map for any Hamiltonian action of a *torus group* (i.e., a finite Cartesian product of circles) is a convex polytope. This theorem was first published in 1982 independently by Sir Michael Atiyah [At] and Victor Guillemin and Shlomo Sternberg [GS82]. For an up-to-date exposition, see McDuff and Salamon [MS, Section 5.5]. The interplay between the combinatorics and geometry of the polytopes and the topology and geometry of the manifold is a vast and beautiful terrain for mathematical exploration.

To go into more detail on these topics is beyond the scope of this book. We hope that the story of the two-body problem has whetted the appetite of some and satisfied the curiosity of others. We encourage the former to consult the recommended reading and we thank the latter wholeheartedly for their attention.

Recommended Reading

Mathematical Background

Readers wishing to brush up on the multivariable calculus used in these notes may want to consult the following:

> *Basic Multivariable Calculus*, by J. E. Marsden, A. J. Tromba and A. Weinstein, Springer-Verlag, New York, 1993,

or the more sophisticated,

> *Vector Calculus, Linear Algebra and Differential Forms: A Unified Approach*, by J. H. Hubbard and B. B. Hubbard, Prentice Hall, Upper Saddle River, New Jersey, 1999.

Another, classic, source for multivariable calculus as well as well as differential equations, with a brief elementary section on planetary motion is

> *Calculus, Volumes I and II, Second Edition* by T. M. Apostol; Blaisdell, New York, 1969.

An excellent source that unifies mathematical and physical ideas is

> *A Course in Mathematics for Students of Physics, Volumes I and II*, by P. Bamberg and S. Sternberg, Cambridge University Press, Cambridge, 1988.

Differential Geometry

Some differential geometry books limit themselves to submanifolds of Euclidean space. A clear, easy-to-read advanced undergraduate level textbook with an introduction to differential manifolds and differential forms is

> *Analysis on Manifolds*, by J. R. Munkres, Addison-Wesley, Reading, MA, 1991.

Slightly dry but also very reliable is

> *Differential Geometry of Curves and Surfaces*, by M. P. do Carmo, Prentice-Hall, Englewood Cliffs, NJ, 1976.

Others develop the theory from a more general point of view. For an abstract development of the theory one can read

> *Riemannnian Geometry*, by M. P. do Carmo, Birkhäuser, Boston, 1992.

Three books that bring physics into the discussion explicitly are

> *Topology, Geometry, and Gauge Fields: Foundations*, by Gregory L. Naber, Springer-Verlag, Texts in Applied Mathematics 25, New York, 1997,

a solid mathematical text that gives careful exposition of the relevance of the math to the physics;

> *The Geometry of Physics: An Introduction*, by Theodore Frankel, Cambridge University Press, Cambridge, England, 1997,

a crash course in the advanced mathematics underpinning much of modern mathematical physics; and

> Burke, William L. *Spacetime Geometry and Cosmology*, by William L. Burke, Univ. Sci. Books, Mill Valley CA,

which discusses the geometry of one-forms explicitly.

Symplectic Geometry

Two senior-to-graduate level books treating all of the mathematical topics in these notes are:

"An Introduction to Lie Groups and Symplectic Geometry," by R. L. Bryant, in *Geometry and Quantum Field Theory*, D. S. Freed and K. K. Uhlenbeck, eds., American Mathematical Society (IAS/Park City mathematics series, volume I), Providence, 1995,

and

Applications of Lie Groups to Differential Equations, Second Edition, by P. J. Olver, Springer-Verlag, New York, 1993.

Slightly more sophisticated are

Introduction to Symplectic Topology, by D. McDuff and D. Salamon, Clarendon Press, Oxford, 1995,

and

"Symplectic Geometry," by V. I. Arnold and A. B. Givental, in: *Encyclopaedia of Mathematical Sciences*, Volume 4, V. I. Arnold and S. P. Novikov, eds., Springer-Verlag, Berlin, 1990.

Advanced discussions of the symmetries of the two-body problem and related topics are given in

Global Aspects of Classical Integrable Systems, by R. H. Cushman and L. M. Bates, Birkhäuser Verlag, Basel, 1991

and

Variations on a Theme of Kepler, by V. Guillemin and S. Sternberg, American Mathematical Society Colloquium Publications, Vol. 42, American Mathematical Society, Providence, RI, 1990.

Classical Mechanics

Serious students of mechanics should familiarize themselves with the modern classics, the easiest of which is

Classical Mechanics, by H. Goldstein, Addison-Wesley, Reading, MA, 1950.

Readers who are ready to use more elbow grease will find sophisticated, beautiful, modern treatments of the most important topics in mathematical mechanics, from the most basic to the esoteric, in

Mathematical Methods of Classical Mechanics, Second Edition, by
V. I. Arnold, Springer-Verlag, New York, 1989.

A comprehensive, concise development of the mathematical theory underlying
mechanics, covering all the topics discussed in these lecture notes (and much
more) in efficient notation is

Foundations of Mechanics, Second Edition, by R. Abraham and J.
Marsden, Addison-Wesley, Reading, MA, 1978.

Celestial Mechanics and Gravitation

The proofs in Newton's *Principia* are not easily accessible, but the introductory
sections, "Definitions" and "Axioms, or Laws of Motion," are beautifully clear.
A recent translation is

The Principia : Mathematical Principles of Natural Philosophy by
Isaac Newton, I. Bernard Cohen (Translator), Anne Whitman (Translator), Univ. California Press, Los Angeles, 1999.

The formidable task of reading some of the proofs in *Principia* has been
made much more pleasant for the modern reader by the commentary and rewrit-
ings in

Newton's Principia for the Common Reader, S. Chandrasekhar,
Clarendon Press, Oxford, 1995.

An introductory-level physics treatment of gravitation and planetary motion,
including an elementary derivation of the inverse-square force law from the
assumption that orbits are circular, and the justification for treating celestial
bodies like point masses, can be found in Chapter 11 of

Physics, by M. Alonso and E. J. Finn, Addison Wesley, Wokingham,
England, 1992.

Johannes Kepler was an exuberant and unusual character. The novelist Arthur
Koestler has written an amusing and compelling biography which portrays a
dogged, obsessive quest for geometric order in the heavens:

The Watershed: A Biography of Johannes Kepler, by Arthur Koestler,
Doubleday & Co., Garden City, New York, 1960.

A more technical, yet moving and easy-to-read account of Kepler's extraction of his laws from the earthbound data collected by Tycho Brahe is available in Chapter 7 of

> *Project Physics, Text* by G. Holton, F. J. Rutherford and F. G. Watson, 1981.

At the advanced undergraduate level, physicists' treatments of the two-body problem, Hamiltonians and other related problems appear in

> *Newtonian Dynamics*, by R. Baierlin, McGraw Hill, Inc., New York, 1983,

and

> *Classical Dynamics of Particles and Systems, Fourth Edition*, by J. B. Marion and S. T. Thornton, Saunders College Publishing, Fort Worth, 1995.

An especially charming book at this level, including an ebullient introduction and an annotated list of major reference books, including many original sources, is:

> *Adventures in Celestial Mechanics: A First Course in the Theory of Orbits*, by V. G. Szebehely, University of Texas Press, Austin, 1989.

There is even more symmetry in the two-body problem than we have described in this book. The associated conserved quantities are the axis and eccentricity of the ellipse, encoded in the Runge-Lenz vector. A lovely elementary treatment (in French) is available in

> "Les origines du calcul symplectique chez Lagrange," by P. Iglesias, in: *Le journal de maths des elèves*, Volume 1 (1995), No. 3, pp. 153–161.

Solutions

Contents

0 Preliminaries

Exercise 2 *Show that for any $m \times n$ matrix A and and $n \times m$ matrix B we have $\mathrm{tr}(AB) = \mathrm{tr}(BA)$. Also show that if A and B are matrix-valued functions of t then $\frac{d}{dt}\mathrm{tr}(AB) = \mathrm{tr}((\frac{d}{dt}A)B) + \mathrm{tr}(A\frac{d}{dt}B)$, where $\frac{d}{dt}$ acts entry by entry on matrices.*

Solution

Let A be a $m \times n$ matrix and B be a $n \times m$ matrix. Let a_{ij}, b_{ij}, $(ab)_{ij}$, and $(ba)_{ij}$ denote the entries of A, B, AB, and BA, respectively. Then we know that

$$(ab)_{ii} = \sum_{j=1}^{n} a_{ji}b_{ij} \text{ and } (ba)_{jj} = \sum_{i=1}^{m} b_{ij}a_{ji}.$$

So

$$tr(AB) = \sum_{i=1}^{m}(ab)_{ii} = \sum_{i=1}^{m}\sum_{j=1}^{n} a_{ji}b_{ij}$$

$$tr(BA) = \sum_{j=1}^{n}(ba)_{jj} = \sum_{j=1}^{n}\sum_{i=1}^{m} b_{ij}a_{ji},$$

which proves that $tr(AB) = tr(BA)$. On the other hand,

$$\frac{d}{dt}tr(AB) = \sum_{i=1}^{m}\sum_{j=1}^{n} \frac{d}{dt}(a_{ji}b_{ij})$$

$$= \sum_{i=1}^{m}\sum_{j=1}^{n}\left(\frac{d}{dt}a_{ji}\right)b_{ij} + a_{ji}\frac{d}{dt}b_{ji} = tr\left(\frac{d}{dt}A\,B\right) + tr\left(A\frac{d}{dt}B\right).$$

1   The Two-Body Problem

Exercise 5 *Check that Kepler's three laws imply that the ellipse on which the planet moves, as well as the position of the planet at any time, is determined by the initial position and velocity of the planet. There is a solution requiring only the geometry of the ellipse and a smidgen of differential calculus.*

Solution

We will use several properties of ellipses:

1. An ellipse is determined by the positions of its foci and the length a of the semimajor axis.

2. For any point p on the ellipse, the sum of the distances from p to each of the foci is equal to twice the length of the semimajor axis.

3. If one reflects the ray from one focus to a point on the ellipse in the ellipse (more specifically, in the tangent line to the ellipse at the point in question) the resulting ray passes through the other focus of the ellipse.

Let s denote the constant position of the sun, let p denote the initial position of the planet and let $\mathbf{v}$ denote the initial velocity of the planet. Because p and $\mathbf{v}$ together determine the tangent line to the ellipse at p, we know by Property 3 that the other focus of the ellipse lies on a certain line. It follows from Property 2 that the position of the other focus will be determined by the length a of the semimajor axis. So by Property 1 we will have determined the ellipse if we can find the length a of the semimajor axis.

We will find the length a of the semimajor axis by equating two different expressions for the area A of the ellipse. To get the first expression we use the formula $A = \pi ab$, where b denotes the length of the semiminor axis of the ellipse. Using the geometry of the ellipse, Property 2 and the law of cosines we can write b in terms of the semimajor axis a, the initial distance r between the sun and the planet and the angle ϕ formed by the initial velocity $\mathbf{v}$ and the line from the sun to the initial position of the planet. The result (after a bit of calculation) is $b = \sqrt{r(a - \frac{1}{2}r)(1 + \cos\phi)}$. Note that for any point on any true ellipse we have $a > \frac{1}{2}r$ and $\phi \neq \pi$, so the quantity under the square root is strictly positive. So we conclude that the area $A = \pi a\sqrt{r(a - \frac{1}{2}r)(1 + \cos\phi)}$.

On the other hand, one of Kepler's laws declares that the line from the sun to the planet sweeps out equal areas in equal times. So, because we can calculate the rate of area swept out in unit time, we can obtain another expression for the area of the ellipse. More exactly, the area of the triangle determined by the line from the sun to the planet and the velocity vector is equal to the rate of area swept out in unit time. Some geometry and trigonometry yield the formula $\frac{1}{4}r|\mathbf{v}|(1 + \cos\phi)$ for the area of this triangle. So the total area swept out at any time t is $\frac{1}{4}r|\mathbf{v}|(1 + \cos\phi)t$. By another of Kepler's laws, the period τ of the planet's motion is related to the length a of the semimajor axis by the formula $\tau^2 = Ka^3$, where K is a fixed number. So the area of the ellipse is $\frac{1}{4}r|\mathbf{v}|(1 + \cos\phi)\sqrt{Ka^3}$.

Squaring both expressions for the area, equating the results, and simplifying we have

$$\pi^2(a - \frac{1}{2}r) = \frac{1}{8}r|\mathbf{v}|^2 K^2 a.$$

Because this is a linear equation in a, there is at most one solution. But we know a priori that there is at least one solution since our planet travels on an ellipse. So there is one unique solution a; hence by the reasoning above we have determined the ellipse of motion.

Once the ellipse is known, the position at any given time is determined by the area swept out in that time, which we have already calculated. So we have determined the orbit and the position of the planet as a function of time from the initial position and velocity, as desired.

Exercise 6 *(Vector calculus review) Show that for any functions* $\mathbf{f}, \mathbf{g} : \mathbb{R} \to \mathbb{R}^3$,

  1.

$$\frac{d}{dt}(\mathbf{f}(t) \times \mathbf{g}(t)) = \mathbf{f}'(t) \times \mathbf{g}(t) + \mathbf{f}(t) \times \mathbf{g}'(t).$$

  2.

$$\frac{d}{dt}(\mathbf{f}(t) \cdot \mathbf{g}(t)) = \mathbf{f}'(t) \cdot \mathbf{g}(t) + \mathbf{f}(t) \cdot \mathbf{g}'(t).$$

  3. Show that for any vectors $\mathbf{v}$ *and* $\mathbf{w}$ *in* $\mathbb{R}^3$, *the nonnegative scalar* $|\mathbf{v} \times \mathbf{w}|$ *is the area of the parallelogram spanned by* $\mathbf{v}$ *and* $\mathbf{w}$.

Solution

1. For vectors $\mathbf{v}, \mathbf{w} \in \mathbb{R}^3$, $\mathbf{v} = (v_1, v_2, v_3)^T$, $\mathbf{w} = (w_1, w_2, w_3)^T$,

$$\mathbf{v} \times \mathbf{w} = \begin{pmatrix} v_2 w_3 - v_3 w_2 \\ v_3 w_1 - v_1 w_3 \\ v_1 w_2 - v_2 w_1 \end{pmatrix}.$$

Omitting the t's for clarity, we get

$$\begin{aligned}
\frac{d}{dt}(\mathbf{f}(t) \times \mathbf{g}(t)) &= \frac{d}{dt} \begin{pmatrix} f_2 g_3 - f_3 g_2 \\ f_3 g_1 - f_1 g_3 \\ f_1 g_2 - f_2 g_1 \end{pmatrix} \\
&= \begin{pmatrix} f_2' g_3 + f_2 g_3' - f_3' g_2 - f_3 g_2' \\ f_3' g_1 + f_3 g_1' - f_1' g_3 - f_1 g_3' \\ f_1' g_2 + f_1 g_2' - f_2' g_1 - f_2 g_1' \end{pmatrix} \\
&= \begin{pmatrix} f_2' g_3 - f_3' g_2 \\ f_3' g_1 - f_1' g_3 \\ f_1' g_2 - f_2' g_1 \end{pmatrix} + \begin{pmatrix} f_2 g_3' - f_3 g_2' \\ f_3 g_1' - f_1 g_3' \\ f_1 g_2' - f_2 g_1' \end{pmatrix} \\
&= \mathbf{f}'(t) \times \mathbf{g}(t) + \mathbf{f}(t) \times \mathbf{g}'(t).
\end{aligned}$$

2. For vectors $\mathbf{v}, \mathbf{w} \in \mathbb{R}^3$, $\mathbf{v} = (v_1, v_2, v_3)^T$, $\mathbf{w} = (w_1, w_2, w_3)^T$,

$$\mathbf{v} \cdot \mathbf{w} = v_1 w_1 + v_2 w_2 + v_3 w_3.$$

Again omitting the t's, we get

$$\begin{aligned}
\frac{d}{dt}(\mathbf{f}(t) \cdot \mathbf{g}(t)) &= \frac{d}{dt}(f_1 g_1 + f_2 g_2 + f_3 g_3) \\
&= f_1' g_1 + f_1 g_1' + f_2' g_2 + f_2 g_2' + f_3' g_3 + f_3 g_3' \\
&= (f_1' g_1 + f_2' g_2 + f_3' g_3) + (f_1 g_1' + f_2 g_2' + f_3 g_3') \\
&= \mathbf{f}'(t) \cdot \mathbf{g}(t) + \mathbf{f}(t) \cdot \mathbf{g}'(t).
\end{aligned}$$

3. The area of a parallelogram is equal to the magnitude of its base times the magnitude of its height. If we take $\mathbf{v}$ to be the base of the parallelogram spanned by $\mathbf{v}$ and $\mathbf{w}$, then the height is the component $\mathbf{w}_{v\perp}$ of $\mathbf{w}$ which is perpendicular to $\mathbf{v}$. We can find the magnitude of $\mathbf{w}_{v\perp}$ using the Pythagorean theorem and the fact that the magnitude of the component $\mathbf{w}_v$ of $\mathbf{w}$ in the direction of $\mathbf{v}$ is equal to $\frac{\mathbf{v} \cdot \mathbf{w}}{|\mathbf{v}|}$.

Since

$$|\mathbf{w}|^2 = |\mathbf{w}_v|^2 + |\mathbf{w}_{v\perp}|^2$$

$$|\mathbf{w}_{v\perp}|^2 = |\mathbf{w}|^2 - |\mathbf{w}_v|^2 = |\mathbf{w}|^2 - \frac{(\mathbf{v} \cdot \mathbf{w})^2}{|\mathbf{v}|^2},$$

we have

$$\begin{aligned}
(\text{area})^2 &= |\mathbf{v}|^2\left(|\mathbf{w}|^2 - \frac{(\mathbf{v} \cdot \mathbf{w})^2}{|\mathbf{v}|^2}\right) \\
&= |\mathbf{v}|^2|\mathbf{w}|^2 - (\mathbf{v} \cdot \mathbf{w})^2 \\
&= (v_1^2 + v_2^2 + v_3^2)(w_1^2 + w_2^2 + w_3^2) - (v_1 w_1 + v_2 w_2 + v_3 w_3)^2 \\
&= v_2^2 w_3^2 - 2v_2 w_2 v_3 w_3 + v_3^2 w_2^2 + v_3^2 w_1^2 - 2v_1 w_1 v_3 w_3 + v_1^2 w_3^2 + v_1^2 w_2^2 \\
&\quad - 2v_1 w_1 v_2 w_2 + v_2^2 w_1^2 \\
&= (v_2 w_3 - v_3 w_2)^2 + (v_3 w_1 - v_1 w_3)^2 + (v_1 w_2 - v_2 w_1)^2 \\
&= |\mathbf{v} \times \mathbf{w}|^2.
\end{aligned}$$

Exercise 7 *Derive these equations:*

$$2\dot{\rho}\dot{\theta} + \rho\frac{d\dot{\theta}}{dt} = 0$$

$$\frac{d\rho}{dt} - \rho\dot{\theta}^2 + \frac{GM}{\rho^2} = 0$$

$$\frac{d\rho}{dt} = \dot{\rho}$$

$$\frac{d\theta}{dt} = \dot{\theta}$$

Solution

We have

$$\frac{d\mathbf{r}}{dt} = \frac{\mathbf{p}^T}{\mu} \tag{1}$$

$$\frac{d\mathbf{p}^T}{dt} = -\frac{GM\mu}{|\mathbf{r}|^3}\mathbf{r} \tag{2}$$

$$\mathbf{r} = \rho \begin{pmatrix} \cos\theta \\ \sin\theta \\ 0 \end{pmatrix} \tag{3}$$

$$\frac{\mathbf{p}}{\mu} = \dot{\rho}\left(\cos\theta \quad \sin\theta \quad 0\right) + \rho\dot{\theta}\left(-\sin\theta \quad \cos\theta \quad 0\right). \tag{4}$$

Combining (1), (3), and (4), we get

$$\frac{d}{dt}\rho \begin{pmatrix} \cos\theta \\ \sin\theta \\ 0 \end{pmatrix} = \begin{pmatrix} \dot{\rho}\cos\theta - \rho\dot{\theta}\sin\theta \\ \dot{\rho}\sin\theta + \rho\dot{\theta}\cos\theta \\ 0 \end{pmatrix}$$

$$\begin{pmatrix} \frac{d\rho}{dt}\cos\theta - \rho\frac{d\theta}{dt}\sin\theta \\ \frac{d\rho}{dt}\sin\theta + \rho\frac{d\theta}{dt}\cos\theta \\ 0 \end{pmatrix} = \begin{pmatrix} \dot{\rho}\cos\theta - \rho\dot{\theta}\sin\theta \\ \dot{\rho}\sin\theta + \rho\dot{\theta}\cos\theta \\ 0 \end{pmatrix}$$

So

$$\frac{d\rho}{dt} = \dot{\rho}$$

$$\text{and} \quad \frac{d\theta}{dt} = \dot{\theta}.$$

Now, combining (2), (3), and (4), we get

$$-\frac{GM\mu}{\rho^2}\begin{pmatrix}\cos\theta\\\sin\theta\\0\end{pmatrix}=\frac{d}{dt}\mu\begin{pmatrix}\dot\rho\cos\theta-\rho\dot\theta\sin\theta\\\dot\rho\sin\theta+\rho\dot\theta\cos\theta\\0\end{pmatrix}$$

$$-\frac{GM}{\rho^2}\begin{pmatrix}\cos\theta\\\sin\theta\\0\end{pmatrix}=\begin{pmatrix}\dfrac{d\dot\rho}{dt}\cos\theta-\dot\rho\dot\theta\sin\theta-\dot\rho\dot\theta\sin\theta-\rho\dfrac{d\dot\theta}{dt}\sin\theta-\rho\dot\theta^2\cos\theta\\[2mm]\dfrac{d\dot\rho}{dt}\sin\theta+\dot\rho\dot\theta\cos\theta+\dot\rho\dot\theta\cos\theta+\rho\dfrac{d\dot\theta}{dt}\cos\theta-\rho\dot\theta^2\sin\theta\\[2mm]0\end{pmatrix}$$

$$2\dot\rho\dot\theta+\rho\frac{d\dot\theta}{dt}=0$$

and $\dfrac{d\dot\rho}{dt}-\rho\dot\theta^2+\dfrac{GM}{\rho^2}=0.$

Exercise 8 *Derive the equation*

$$\frac{d^2\rho}{dt^2}=-\frac{|\tilde{\mathbf{L}}|^2}{\mu^2}q^2\frac{d^2q}{d\theta^2}.$$

Solution

First we will change the dependent variable, using the substitution $\rho=q^{-1}$.
Then we will change the independent variable, taking advantage of the chain
rule and the fact that

$$\frac{d\theta}{dt}=\frac{|\tilde{\mathbf{L}}|}{\mu\rho^2}=q^2|\tilde{\mathbf{L}}|/\mu.$$

The explicit calculation is:

$$\frac{d^2\rho}{dt^2}=\frac{d}{dt^2}q^{-1}=\frac{d}{dt}(-q^{-2}\frac{dq}{dt})$$

$$=2q^{-3}\left(\frac{dq}{dt}\right)^2-q^{-2}\frac{d^2q}{dt^2}$$

$$=2q^{-3}\left(\frac{dq}{d\theta}\right)^2\left(\frac{d\theta}{dt}\right)^2-q^{-1}\frac{d}{dt}\left(\frac{dq}{d\theta}\frac{d\theta}{dt}\right)$$

$$=2q^{-3}\left(\frac{dq}{d\theta}\right)^2\left(\frac{d\theta}{dt}\right)^2-q^{-2}\frac{d^2q}{d\theta^2}\left(\frac{d\theta}{dt}\right)^2-q^{-2}\frac{dq}{d\theta}\frac{d}{dt}\left(q^2|\tilde{\mathbf{L}}|/\mu\right)$$

$$=2q^{-3}\left(\frac{dq}{d\theta}\right)^2\left(\frac{d\theta}{dt}\right)^2-q^{-2}\frac{d^2q}{d\theta^2}\left(\frac{d\theta}{dt}\right)^2-2q^3\left(\frac{dq}{d\theta}\right)^2\left(\frac{d\theta}{dt}\right)\left(q^2|\tilde{\mathbf{L}}|/\mu\right)$$

$$= -q^{-2}\frac{d^2q}{d\theta^2}\left(\frac{d\theta}{dt}\right)^2$$

$$= -\frac{|\tilde{L}|^2}{\mu^2}q^2\frac{d^2q}{d\theta^2}.$$

2   Phase Spaces are Symplectic Manifolds

Exercise 12 *Fix a nonzero vector* $\mathbf{v} \in \mathbb{R}^2$. *For which vectors* $\mathbf{w}$ *do we have* $A(\mathbf{v}, \mathbf{w}) = 1$? *Draw* $\mathbf{v}$ *and the set of possible* $\mathbf{w}$'s.

Solution

Fix $\mathbf{v} = (r_1, p_1)^T$ and let $\mathbf{w} = (r_2, p_2)^T$ be an arbitrary vector in $\mathbb{R}^2$. The equation

$$A(\mathbf{v}, \mathbf{w}) = r_1 p_2 - r_2 p_1 = 1$$

defines a line in $\mathbb{R}^2$ which is parallel to $\mathbf{v}$ and contains the vector $(\frac{-1}{2p_1}, \frac{1}{2r_1})^T$. This makes sense because the area of a parallelogram is equal to its base times its height. All the points on the line described by the above equation are $\frac{1}{|\mathbf{v}|}$ units from the line containing $\mathbf{v}$; hence any parallelogram spanned by $\mathbf{v}$ and $\mathbf{w}$ with $\mathbf{w}$ in the line we described will have base $|\mathbf{v}|$, height $\frac{1}{|\mathbf{v}|}$ and area 1.

Exercise 14 *The product* $\mathbf{u} \cdot (\mathbf{v} \times \mathbf{w})$ *is called the* triple scalar product *of* $\mathbf{u}$, $\mathbf{v}$ *and* $\mathbf{w}$.

1. *Show that* $\mathbf{u} \cdot (\mathbf{v} \times \mathbf{w})$ *is the determinant of the matrix with columns* $\mathbf{u}$, $\mathbf{v}$ *and* $\mathbf{w}$.

2. *Interpret* $\mathbf{u} \cdot (\mathbf{v} \times \mathbf{w})$ *geometrically in terms of the parallelipiped spanned by* $\mathbf{u}$, $\mathbf{v}$ *and* $\mathbf{w}$.

3. *Show that* $\mathbf{u} \cdot (\mathbf{v} \times \mathbf{w}) = \mathbf{v} \cdot (\mathbf{w} \times \mathbf{u}) = \mathbf{w} \cdot (\mathbf{u} \times \mathbf{v})$ *for all vectors* $\mathbf{u}$, $\mathbf{v}$, $\mathbf{w} \in \mathbb{R}^3$, *by direct computation or by using 1 or 2.*

4. *Show that a* 3×3 *matrix M has determinant 1 if and only if the function* $\mathbb{R}^3 \to \mathbb{R}^3$, $\mathbf{v} \mapsto M\mathbf{v}$ *takes the unit cube in the domain to a parallelepiped of signed volume 1.*

Solution

1. This is easy to see because the cross product $\mathbf{v} \times \mathbf{w}$ is sometimes interpreted as the "determinant" of the matrix

$$\begin{pmatrix} \partial_1 & v_1 & w_1 \\ \partial_2 & v_2 & w_2 \\ \partial_3 & v_3 & w_3 \end{pmatrix},$$

where ∂_1, ∂_2 and ∂_3 are the standard basis vectors in $\mathbb{R}^3$. However, the dot product $\mathbf{u} \cdot (\mathbf{v} \times \mathbf{w})$ is equal to the sum over $i = 1, 2, 3$ of u_i times the component of $\mathbf{v} \times \mathbf{w}$ in the direction of ∂_i. So we can replace the ∂_i's in the previous matrix with u_i's to get

$$u \cdot (v \times w) = \det \begin{pmatrix} u_1 & v_1 & w_1 \\ u_2 & v_2 & w_2 \\ u_3 & v_3 & w_3 \end{pmatrix}.$$

3. Using the result from question 1, we can say that

$$
\begin{aligned}
v \cdot (w \times u) &= \det \begin{pmatrix} v_1 & w_1 & u_1 \\ v_2 & w_2 & u_2 \\ v_3 & w_3 & u_3 \end{pmatrix} \\
&= \det \left[\begin{pmatrix} u_1 & v_1 & w_1 \\ u_2 & v_2 & w_2 \\ u_3 & v_3 & w_3 \end{pmatrix} \begin{pmatrix} 0 & 0 & 1 \\ 1 & 0 & 0 \\ 0 & 1 & 0 \end{pmatrix} \right] \\
&= \det \begin{pmatrix} u_1 & v_1 & w_1 \\ u_2 & v_2 & w_2 \\ u_3 & v_3 & w_3 \end{pmatrix} \det \begin{pmatrix} 0 & 0 & 1 \\ 1 & 0 & 0 \\ 0 & 1 & 0 \end{pmatrix} \\
&= \det \begin{pmatrix} u_1 & v_1 & w_1 \\ u_2 & v_2 & w_2 \\ u_3 & v_3 & w_3 \end{pmatrix} \quad (1) \\
&= u \cdot (v \times w).
\end{aligned}
$$

The proof that $w \cdot (u \times v) = u \cdot (v \times w)$ is similar.

2. The value of $\mathbf{u} \cdot (\mathbf{v} \times \mathbf{w})$ is equal to the signed volume of the parallelepiped spanned by $\mathbf{u}$, $\mathbf{v}$, and $\mathbf{w}$. To see this, view the parallelogram spanned by $\mathbf{v}$ and $\mathbf{w}$ as the base of the parallelepiped. The area of the base is equal to $|\mathbf{v} \times \mathbf{w}|$. Recall that $\mathbf{v} \times \mathbf{w}$ is perpendicular to the plane containing $\mathbf{v}$ and $\mathbf{w}$. So

the dot product $\mathbf{u} \cdot \frac{\mathbf{v} \times \mathbf{w}}{|\mathbf{v} \times \mathbf{w}|}$ is the component of $\mathbf{u}$ perpendicular to the base parallelogram, which can be thought of as the signed height of the parallelepiped. The signed volume is therefore

$$\mathbf{u} \cdot \frac{\mathbf{v} \times \mathbf{w}}{|\mathbf{v} \times \mathbf{w}|} |\mathbf{v} \times \mathbf{w}| = \mathbf{u} \cdot (\mathbf{v} \times \mathbf{w}).$$

4. We have shown that the signed volume of the parallelepiped spanned by $\mathbf{u}$, $\mathbf{v}$, and $\mathbf{w}$ is equal to

$$\det \begin{pmatrix} u_1 & v_1 & w_1 \\ u_2 & v_2 & w_2 \\ u_3 & v_3 & w_3 \end{pmatrix}.$$

So the signed volume of the parallelepiped spanned by A applied to the standard basis vectors is

$$\det \left(A \begin{pmatrix} 1 \\ 0 \\ 0 \end{pmatrix} \middle| A \begin{pmatrix} 0 \\ 1 \\ 0 \end{pmatrix} \middle| A \begin{pmatrix} 0 \\ 0 \\ 1 \end{pmatrix} \right)$$

$$= \det \left[A \begin{pmatrix} 1 & 0 & 0 \\ 0 & 1 & 0 \\ 0 & 0 & 1 \end{pmatrix} \right]$$

$$= \det A.$$

Hence the signed volume of the unit cube is preserved if and only if $\det A = 1$.

Exercise 18 *Show that there is a one-to-one correspondence between antisymmetric bilinear forms on $\mathbb{R}^2$ and constant, antisymmetric 2×2 matrices.*

Solution

It turns out that given a basis v, w for $\mathbb{R}^2$, any antisymmetric bilinear form on $\mathbb{R}^2$ is completely determined by its value on the ordered pair (v, w). Since, if F is an antisymmetric bilinear form on $\mathbb{R}^2$, then

$$F(av + bw, cv + dw) = ac\, F(v, v) + ad\, F(v, w)$$
$$+ bc\, F(w, v) + bd\, F(w, w) \quad \text{by linearity}$$
$$= (ad - bc)F(v, w) \quad \text{by antisymmetry}$$

So F can be identified with the matrix

$$\begin{pmatrix} 0 & F(v, w) \\ -F(v, w) & 0 \end{pmatrix}.$$

Conversely, given an antisymmetric matrix A, one can construct an antisymmetric bilinear form F on $\mathbb{R}^2$ by

$$F(v, w) = v^T A w.$$

The fact that F is bilinear comes directly from the linearity of matrix multiplication. To see that F is antisymmetric, note that

$$\begin{aligned} F(w, v) &= w^T A v \\ &= (A^T w)^T v \\ &= (-Aw)^T v \\ &= (v^T (-A) w)^T \\ &= -(v^T A w)^T \\ &= -F(v, w)^T. \end{aligned}$$

However, $F(v, w)$ is a scalar, so $F(v, w)^T = F(v, w)$. Hence $F(w, v) = -F(v, w)$. The reader can check that these two identifications (getting a matrix from a two-form and getting a two-form from a matrix) are invertible and hence define a one-to-one correspondence between antisymmetric bilinear forms on $\mathbb{R}^2$ and antisymmetric 2×2 matrices.

Exercise 20 *Show that every antisymmetric bilinear form on $\mathbb{R}^2$ can be written as a wedge product of two covectors.*

Solution

It was shown in the solution to Exercise 18 that given any antisymmetric bilinear two-form F on $\mathbb{R}^2$ and any basis v, w for $\mathbb{R}^2$, the value of F on any two vectors is given by

$$F(av + bw, cv + dw) = (ad - bc)F(v, w).$$

Given that relation, it is easy to verify that F can be written as $\alpha \wedge \beta$, where

$$\alpha(av + bw) = aF(v, w)$$
$$\beta(av + bw) = b.$$

Recall that $F(v, w)$ is just a constant; hence α and β are well defined.

Exercise 24 *Show that there is a one-to-one correspondence between antisymmetric bilinear forms on $\mathbb{R}^n$ and constant, antisymmetric $n \times n$ matrices.*

Solution

First we will show that, given a basis $b_1, \ldots, b_n$ of $\mathbb{R}^n$, any bilinear form F on $\mathbb{R}^n$ (not necessarily antisymmetric) can be written as $F(v, w) = v^T A w$, where A is an $n \times n$ matrix with entry a_{ij} equal to $F(b_i, b_j)$. We can see this by saying that if

$$v = \sum_{i=1}^n v_i b_i \quad \text{and} \quad w = \sum_{j=1}^n w_j b_j,$$

then, by bilinearity,

$$F(v, w) = \sum_{i=1}^n \sum_{j=1}^n v_i w_j F(b_i, b_j)$$

which can be shown by matrix computation to be equal to $v^T A w$, with A defined as above. Note that A^T has entry a_{ij} equal to $F(b_j, b_i)$. Hence, if F is antisymmetric, we have $A^T = -A$, meaning A is antisymmetric. Conversely, given an antisymmetric $n \times n$ matrix A, we can define a form F by $F(v, w) = v^T A w$. See the solution to Exercise 18 for the proof that the resulting F is bilinear and antisymmetic.

Exercise 26 *Show that, if we define the vector $\frac{\partial}{\partial \theta}$ to be $\frac{\partial}{\partial \theta}(\rho \cos \theta, \rho \sin \theta)^T$ then*

$$\frac{\partial}{\partial \theta} = \begin{pmatrix} -p \\ r \end{pmatrix} = -p \frac{\partial}{\partial r} + r \frac{\partial}{\partial p}.$$

Similarly show that

$$\frac{\partial}{\partial \rho} = \begin{pmatrix} \frac{r}{\sqrt{r^2+p^2}} \\ \frac{p}{\sqrt{r^2+p^2}} \end{pmatrix} = \frac{r}{\sqrt{r^2 + p^2}} \frac{\partial}{\partial r} + \frac{p}{\sqrt{r^2 + p^2}} \frac{\partial}{\partial p}.$$

Finally, use the chain rule to check that these equalities make sense for partial differentiation.

Solution

For the first part, recall that $\frac{\partial}{\partial r} = (1, 0)^T$ and $\frac{\partial}{\partial p} = (0, 1)^T$ and calculate:

$$\frac{\partial}{\partial \theta} := \frac{\partial}{\partial \theta} \left(\rho \cos\theta, \rho \sin\theta \right) = \left(-\rho \sin\theta, \rho \cos\theta \right) = \left(-p, r \right)$$

$$= -p\frac{\partial}{\partial r} + r\frac{\partial}{\partial p}.$$

Next, for any differentiable function $f : \mathbb{R}^2 \to \mathbb{R}$, $(r, p)^T \mapsto f(r, p)$ we have

$$\frac{\partial f}{\partial \theta} = \frac{\partial}{\partial \theta} f\left(\rho \cos\theta, \rho \sin\theta \right) = \rho\left(\cos\theta \frac{\partial f}{\partial p} - \sin\theta \frac{\partial f}{\partial r} \right).$$

Similar calculations yield the desired result for $\frac{\partial}{\partial \rho}$.

Exercise 29 *Let $f : \mathbb{R}^3 \to \mathbb{R}^3$ be a differentiable function. Show that the function $\mathrm{tr}(J_f) : \mathbb{R}^3 \to \mathbb{R}$ is equal to the divergence of the vector field f.*

Solution

Using the usual basis for $\mathbb{R}^3$ we can define coordinate functions by

$$\begin{pmatrix} f_1(x, y, z) \\ f_2(x, y, z) \\ f_3(x, y, z) \end{pmatrix} := f \begin{pmatrix} x \\ y \\ z \end{pmatrix}.$$

By the definition of the divergence, we have

$$\mathrm{div}\, f = \frac{\partial f_1}{\partial x} + \frac{\partial f_2}{\partial y} + \frac{\partial f_3}{\partial z}.$$

On the other hand,

$$\mathrm{tr} J_f = \mathrm{tr} \begin{pmatrix} \dfrac{\partial f_1}{\partial x} & \dfrac{\partial f_1}{\partial y} & \dfrac{\partial f_1}{\partial z} \\[2mm] \dfrac{\partial f_2}{\partial x} & \dfrac{\partial f_2}{\partial y} & \dfrac{\partial f_2}{\partial z} \\[2mm] \dfrac{\partial f_3}{\partial x} & \dfrac{\partial f_3}{\partial y} & \dfrac{\partial f_3}{\partial z} \end{pmatrix} = \frac{\partial f_1}{\partial x} + \frac{\partial f_2}{\partial y} + \frac{\partial f_3}{\partial z}.$$

So the two functions are equal.

Exercise 33 *Let f be the distance function in Cartesian coordinates. Write a formula for f and calculate $\Gamma^* f$ explicitly. Your answer should be the function $(\rho, \theta)^T \mapsto \rho$.*

Solution

The distance function in Cartesian coordinates is given by

$$f = \sqrt{x^2 + y^2}.$$

Pulling this back by the map

$$\Gamma : \mathbb{R}^+ \times \mathbb{R} \to \mathbb{R}^2 \backslash (0, 0)^T$$
$$\begin{pmatrix} \rho \\ \theta \end{pmatrix} \mapsto \begin{pmatrix} \rho \cos \theta \\ \rho \sin \theta \end{pmatrix},$$

we get

$$\begin{aligned} f &= \sqrt{(\rho \cos \theta)^2 + (\rho \sin \theta)^2} \\ &= \sqrt{\rho^2 (\cos^2 \theta + \sin^2 \theta} \\ &= \sqrt{\rho^2} \\ &= |\rho| = \rho. \end{aligned}$$

We have therefore obtained the desired map

$$f : \begin{pmatrix} \rho \\ \theta \end{pmatrix} \mapsto \rho.$$

3   Differential Geometry

Exercise 37 *Consider the functions* $[0, 2\pi) \to \mathbb{R}$, $t \mapsto \sin t$ *and* $[0, 2\pi) \to \mathbb{R}$, $t \mapsto \sin \frac{t}{2}$. *First check that both functions are differentiable on* $(0, 2\pi)$ *and have one-sided derivatives at* 0. *Next, consider the corresponding functions on the circle* $S^1 \to \mathbb{R}$, $\begin{pmatrix} x \\ y \end{pmatrix} \mapsto \sin \left(\tilde{\gamma}^{-1} \begin{pmatrix} x \\ y \end{pmatrix} \right)$ *and* $S^1 \to \mathbb{R}$, $\begin{pmatrix} x \\ y \end{pmatrix} \mapsto \sin \left(\frac{1}{2} \tilde{\gamma}^{-1} \begin{pmatrix} x \\ y \end{pmatrix} \right)$ *and argue informally that the first is differentiable at* $\begin{pmatrix} 1 \\ 0 \end{pmatrix}$ *while the second is not. (Note that you cannot argue formally without a suitable definition of differentiability of a function on the circle.)*

Solution

It is not hard to check that both functions are differentiable on $(0, \pi)$, since both can be extended to functions on $\mathbb{R}$ whose derivatives are well known: the derivative of $\sin t$ is $\cos t$, and the derivative of $\sin \frac{t}{2}$ is $\frac{1}{2} \cos \frac{t}{2}$. We see from these formulas that the one-sided derivatives of these functions at 0 are, respectively, 1 and $\frac{1}{2}$.

To see whether the corresponding functions on the circle are differentiable at the point $\begin{pmatrix} 1 \\ 0 \end{pmatrix}$ we need to check if the values of the functions and their derivatives near 2π match the values at 0. Let us look at the first function. Its value at 0 is 0, and its value $\sin t$ approaches 0 as t approaches 2π. Likewise, its one-sided derivative at 0 is 1, and its derivative $\cos t$ approaches 1 as t approaches 2π. So the first function is differentiable on the circle. Next consider the second function. Its value at 0 is 0, and its value $\sin \frac{t}{2}$ approaches 0 as t approaches 2π. On the other hand, its one-sided derivative at 0 is $\frac{1}{2}$, but its derivative $\frac{1}{2} \cos \frac{t}{2}$ approaches $-\frac{1}{2}$ as t approaches 2π. So the second function is not differentiable on the circle, because its derivatives do not match up as you approach the point $\begin{pmatrix} 1 \\ 0 \end{pmatrix}$ from the two different sides.

Exercise 41 *Check that the coordinate system defined by γ_3 is compatible with the coordinate system defined by γ_1.*

Solution

We must look at the change-of-coordinates function on the overlap and argue that this function and its inverse are infinitely differentiable. First we find the overlap: the range of γ_1 includes everything but the north and south poles and the international date line. The range of γ_3 is the whole northern hemisphere, not including the equator. It follows that the domain of the change of coordinates function $\gamma_3^{-1} \circ \gamma_1$ is $(0, \frac{\pi}{2}) \times (-\pi, \pi)$. In addition, we can calculate that

$$\gamma_3^{-1} \circ \gamma_1 \begin{pmatrix} \theta_1 \\ \phi_1 \end{pmatrix} = \begin{pmatrix} \sin \theta_1 \cos \phi_1 \\ \sin \theta_1 \sin \phi_1 \end{pmatrix},$$

which is infinitely differentiable. Explicitly, the Jacobian matrix of this function is

$$\begin{pmatrix} \cos \theta_1 \cos \phi_1 & -\sin \theta_1 \sin \phi_1 \\ \cos \theta_1 \sin \phi_1 & \sin \theta_1 \cos \phi_1 \end{pmatrix},$$

each of whose entries is infinitely differentiable.

To see that the inverse is infinitely differentiable we use the inverse function theorem, which guarantees that wherever the Jacobian matrix of the function is nonsingular, the Jacobian matrix of the inverse equals the inverse matrix of the Jacobian. In our case the determinant of the Jacobian is $\cos\theta_1 \sin\theta_1$, which is infinitely differentiable and nonzero everywhere on the domain since $\theta_1 \in (0, \frac{\pi}{2}$. This, along with Cramer's rule, guarantees that the Jacobian of the inverse has infinitely differentiable entries. Hence the inverse is infinitely differentiable.

Exercise 45 *Show formally that for any natural number k the Euclidean space* $\mathbb{R}^k$ *is a k-manifold.*

Solution

$\mathbb{R}^k$ is a manifold with a one coordinate patch–itself! Formally, we let U be $\mathbb{R}^k$ (which is an open set) and define

$$\gamma : U \to \mathbb{R}^k$$

to be the identity map on $\mathbb{R}^k$, i.e. $\gamma(x) = x$ for all $x \in \mathbb{R}^k$. Obviously $\gamma(U) = \mathbb{R}^k$, so condition 2 is satisfied, and condition 3 is vacuous since there are no overlaps of patches.

Exercise 48 *Show that if M is an m-dimensional manifold and N is an n-dimensional manifold, then the Cartesian product M × N has the structure of an (m + n)-dimensional manifold.*

Solution

Suppose M has coordinate patches $\gamma_i : U_i \to M$ and N has coordinate patches $\delta_j : V_j \to N$ with U_i open in $\mathbb{R}^m$ and V_j open in $\mathbb{R}^n$. We form coordinate patches on $M \times N$ by taking

$$\gamma_i \times \delta_j : U_i \times V_j \to M \times N$$
$$(u, v) \mapsto (\gamma_i(u), \delta_j(v)).$$

Condition 2 is easily verified as follows: for every $(x, y) \in M \times N$, we know that there is some i such that $x \in \gamma_i(U_i)$ and likewise some j such that $y \in \delta_j(V_j)$. We therefore know that $(x, y) \in \gamma_i \times \delta_j(U_i \times V_j)$. To verify condition 3, take two coordinate patches

$$\gamma_i \times \delta_j : U_i \times V_j \to M \times N$$
$$\gamma_k \times \delta_\ell : U_k \times V_\ell \to M \times N$$

which overlap on

$$W = \gamma_i \times \delta_j(U_i \times V_j) \cap \gamma_k \times \delta_\ell(U_k \times V_\ell).$$

We must verify that $(\gamma_i \times \delta_j)^{-1}(W)$ and $(\gamma_k \times \delta_\ell)^{-1}(W)$ are open, and that

$$\Gamma := (\gamma_k \times \delta_\ell)^{-1} \circ (\gamma_i \times \delta_j) : (\gamma_i \times \delta_j)^{-1}(W) \to (\gamma_k \times \delta_\ell)^{-1}(W)$$

is infinitely differentiable. To do this, the reader must convince him or herself that, given invertible functions a, b, c, and d with appropriate ranges and domains so that the compositions $a \circ c$ and $b \circ d$ make sense, the following equations hold:

$$(a \times b)^{-1} = a^{-1} \times b^{-1}$$
$$(a \times b) \circ (c \times d) = (a \circ c) \times (b \circ d).$$

Then we can see that $(\gamma_i \times \delta_j)^{-1}(W) = \gamma_i^{-1}(W) \times \delta_j^{-1}(W)$ is the Cartesian product of two open sets, which is open. Similarly, $(\gamma_k \times \delta_\ell)^{-1}(W)$ is open. As for the transition function,

$$\Gamma = (\gamma_k \times \delta_\ell)^{-1} \circ (\gamma_i \times \delta_j) = (\gamma_k^{-1} \circ \gamma_i) \times (\delta_\ell^{-1} \circ \delta_j),$$

which is the Cartesian product of two infinitely differentiable functions, making Γ infinitely differentiable.

Exercise 49 *Show that the second function defined in Exercise 37 is not infinitely differentiable.*

Solution

Consider the function $f : S^1 \to \mathbb{R}$ defined by $(x, y)^T \mapsto \sin(\frac{1}{2}\tilde{\gamma}^{-1}(x, y)$, where $\tilde{\gamma} : [0, 2\pi) \to S^1$ is defined by $t \mapsto (\cos t, \sin t)$. Using the definition in the text of γ_2 we have

$$f \circ \gamma_2(t) = \sin(\frac{1}{2}\tilde{\gamma}^{-1}(\cos t, \sin t) = \begin{cases} \sin(\frac{1}{2}(t + 2\pi)) & t \in (-\pi, 0) \\ \sin(\frac{1}{2}(t + 2\pi)) & t \in [0, \pi). \end{cases}$$

Note that the function $f \circ \gamma_2$ is not infinitely differentiable at $t = 0$ since

$$\lim_{t \to 0^-} \frac{f \circ \gamma_2(t) - f \circ \gamma_2(0)}{t} = -\frac{1}{2} \neq \frac{1}{2} = \lim_{t \to 0^+} \frac{f \circ \gamma_2(t) - f \circ \gamma_2(0)}{t}.$$

Exercise 53 (first half) *Show that every differential two-form on every open set $U \in \mathbb{R}^n$ can be written as a sum of terms of the form*

$$f \, dg \wedge dh,$$

where f, g and h are real-valued functions on U.

Solution

First we will show that every antisymmetric bilinear form F on $\mathbb{R}^n$ can be written as a finite linear combination of wedge products of covectors. Recall from the solution to Exercise 24 that the value of F on vectors v and w can be expressed as

$$F(v, w) = \sum_{i=1}^{n} \sum_{j=1}^{n} v_i w_j F(b_i, b_j)$$

where $b_1, \ldots, b_n$ is a basis for $\mathbb{R}^n$ and v_i and w_j are the ith and jth coordinates of v and w, respectively. Using the antisymmetry of F, we can rewrite this expression as

$$F(v, w) = \sum_{1 \le i < j \le n} (v_i w_j - v_j w_i) F(b_i, b_j).$$

But $v_i w_j - v_j w_i$ is equal to $db_i \wedge db_j(v, w)$. This gives us

$$F(v, w) = \sum_{1 \le i < j \le n} F(b_i, b_j) db_i \wedge db_j(v, w).$$

Now, going back to the problem of a two-form α on $U \in \mathbb{R}^n$, we just replace $F(b_i, b_j)$ with $\alpha(b_i, b_j)$, which is a real-valued function on U. For each i, $1 \le i \le n$, let c_i denote the $i$the coordinate function on U (in the $\{b_i\}$ coordinates). Then $dc_i \wedge dc_j(v, w) = (v_i w_j - v_j w_i)$ and we have

$$\alpha(v, w) = \sum_{i=1}^{n} \sum_{j=1}^{n} \alpha(b_i, b_j) db_i \wedge db_j(v, w)$$

in the required form.

Exercise 55 *Show that if M is an n-manifold and ω is a nondegenerate differential two-form on M then n must be even.*

Solution

Let us start with an antisymmetric bilinear form on $\mathbb{R}^n$, with corresponding antisymmetric matrix A. We will show that if n is odd, A must be singular. This will be shown by looking at the eigenvalues of A. Suppose A has a real eigenvalue λ with eigenvector v. Then

$$v^T A v = v^T \lambda v$$
$$= \lambda (v^T v)$$
$$= \lambda |v|^2.$$

But we know that $v^T A v = 0$ for any $v \in \mathbb{R}^n$. Therefore, any real eigenvalue of A must be zero.

Recall that the characteristic polynomial $P = \det(\lambda I - A)$ is a polynomial of degree n with real coefficients. This means that the number of real roots of P must have the same parity (even or odd) as n since all complex roots come in conjugate pairs. Specifically, if n is odd, P must have at least one real root. A must therefore have a real eigenvalue, and by our work above, this eigenvalue must be zero. Thus the equation $\det(\lambda I - A) = 0$ has $\lambda = 0$ as a solution, meaning $\det A = 0$.

Now suppose we have a differential two-form α on an n-manifold M. Pick a point $x \in M$ which is covered by a coordinate patch $\gamma : U \to M$. We can pull α back by γ to a two-form on $U \subset \mathbb{R}^n$. Restricting to $\gamma^{-1}(x)$ we get an antisymmetric bilinear form on $\mathbb{R}^n$ which corresponds to an $n \times n$ matrix A. We have shown that if n is odd, then A is singular, and hence α is degenerate. Therefore, any nondegenerate two-form on an n-manifold must have n even.

Exercise 56 *State a geometric condition on a function f equivalent to the algebraic condition $df = 0$. State a geometric condition on a one-form α equivalent to the algebraic condition $d\alpha = 0$.*

Solution

We have $df = 0$ if and only if f is constant on each connected component of the domain of f. We have $d\alpha = 0$ if and only if, for any smooth closed curve C contained in an open disk contained entirely in the domain of α, the line integral $\int_C \alpha = 0$.

Exercise 57 *Show that every two-form ω on a two-dimensional manifold M is closed.*

Solution

We must show that $d\omega = 0$ in any coordinate patch. Let (x, y) be any system of coordinates defined locally somewhere on M. Because ω is a two-form, $d\omega$ is a three-form. For any three two-vectors a, b and c we have

$$d\omega\Big(a_x \frac{\partial}{\partial x} + a_y \frac{\partial}{\partial y}, b_x \frac{\partial}{\partial x} + b_y \frac{\partial}{\partial y}, c_x \frac{\partial}{\partial x} + c_y \frac{\partial}{\partial y}\Big)$$

$$= a_x b_x c_x d\omega\Big(\frac{\partial}{\partial x}, \frac{\partial}{\partial x}, \frac{\partial}{\partial x}\Big) + a_x b_x c_y d\omega\Big(\frac{\partial}{\partial x}, \frac{\partial}{\partial x}, \frac{\partial}{\partial y}\Big) + \text{ six more terms}.$$

Since $d\omega$ is alternating, each of these terms is zero. So $d\omega = 0$.

4   Total Energy Functions are Hamiltonian Functions

Exercise 60 *Show that the conservation of H (i.e., the fact that $\frac{dH}{dt} = 0$) follows from the equations $\frac{dr}{dt} = \frac{\partial H}{\partial p}$ and $\frac{dp}{dt} = -\frac{\partial H}{\partial r}$.*

Solution

Using the chain rule to compute $\frac{dH}{dt}$, we get

$$\frac{dH}{dt} = \frac{dr}{dt}\frac{\partial H}{\partial r} + \frac{dp}{dt}\frac{\partial H}{\partial p}$$
$$= \frac{dr}{dt}\left(-\frac{dp}{dt}\right) + \frac{dp}{dt}\frac{dr}{dt}$$
$$= 0.$$

Exercise 63 *Use the nondegeneracy of the symplectic form to show that, given any infinitely differentiable real-valued function H on a phase space M with symplectic form ω, there is a unique vector field X_H satisfying $\iota_{X_H}\omega = -dH$.*

Solution

We will work in the general setting of $\mathbb{R}^{2n}$ even though most physical situations involving the Hamiltonian occur when $n = 1$, 2, or 3. The two-form ω can therefore be represented by a $2n \times 2n$ nowhere singular matrix A with coefficients that are real-valued functions on $\mathbb{R}^{2n}$. Equation 4.3 becomes

$$X_H^T A = -dH.$$

We now use the nonsingularity of A to solve for X_H, obtaining

$$X_H^T = -dH\, A^{-1}.$$

Exercise 66 *Derive the formula given in the text for X_H in the case of a magnetic field in $\mathbb{R}^3$.*

Solution

We write

$$X_H = a_x\frac{\partial}{\partial r_x} + a_y\frac{\partial}{\partial r_y} + a_z\frac{\partial}{\partial r_z} + b_x\frac{\partial}{\partial p_x} + b_y\frac{\partial}{\partial p_y} + b_z\frac{\partial}{\partial p_z},$$

where the coefficients (the a's and b's) are yet to be determined. From the equation $-dH = \iota_{X_H}\omega$ and the various given formulas we require

$$-\frac{1}{m}(p_x dp_x + p_y dp_y + p_z dp_z) - dU = \iota_{a_x \frac{\partial}{\partial r_x} + a_y \frac{\partial}{\partial r_y} + a_z \frac{\partial}{\partial r_z} + b_x \frac{\partial}{\partial p_x} + b_y \frac{\partial}{\partial p_y} + b_z \frac{\partial}{\partial p_z}}$$

$$\left(dp_x \wedge dr_x + dp_y \wedge dr_y + dp_z \wedge dr_z\right.$$

$$\left. + q(B_x dr_y \wedge dr_z + B_y dr_z \wedge dr_x + B_z dr_x \wedge dr_y)\right)$$

$$= -a_x dp_x - a_y dp_y - a_z dp_z + b_x dr_x + b_y dr_y + b_z dr_z$$

$$+ q\left((B_y a_z - B_z a_y)dr_x + (B_z a_x - B_x a_z)dr_y + (B_x a_y - B_y a_x)dr_z\right).$$

We conclude that $a_x = \frac{p_x}{m}$, $a_y = \frac{p_y}{m}$ and $a_z = \frac{p_z}{m}$. It follows that $b_x = \frac{q}{m}(B_z p_y - B_y p_z) - \frac{\partial U}{\partial r_x}$, and similar expressions hold for b_y and b_z. This determines X_H, as desired.

Exercise 71 *Show that the equation $X_H H = 0$ is equivalent to the equation $\Gamma_t^* H = H$.*

Solution

First we will start with the equation $X_H H = 0$. We have

$$\frac{d}{dt}\Gamma_t^* H = \frac{d}{dt}(H \circ \Gamma_t) = dH X_H$$

$$= X_H H = 0.$$

Thus, we can solve for $\Gamma_t^* H$ by taking the integral

$$\Gamma_t^* H = H + \int_0^t 0 \, ds$$

$$= H + 0$$

$$= H.$$

Thus we have derived the equation $\Gamma_t^* H = H$ from the equation $X_H H = 0$. The reader can verify that the steps we used are reversible, so the two equations are equivalent.

Exercise 72 *Consider the flow*

$$\Gamma_t : \quad \mathbb{R}^2 \setminus \{0\} \to \mathbb{R}^2 \setminus \{0\}$$

$$\begin{pmatrix} r \\ p \end{pmatrix} \mapsto \begin{pmatrix} re^t \\ pe^t \end{pmatrix}.$$

Find a symplectic form on $\mathbb{R}^2 \setminus \{0\}$ that is preserved under this flow. Find a Hamiltonian function H for which this Γ_t is the Hamiltonian flow.

Solution

The symplectic form

$$\omega = \frac{1}{pr}dp \wedge dr$$

is preserved under Γ_t, since

$$\Gamma_t^*\omega = \Gamma_t^*(\frac{1}{pr}dp \wedge dr)$$

$$= \frac{1}{(pe^t)(re^t)}d(pe^t) \wedge d(re^t)$$

$$= \frac{1}{e^{2t}pr}e^{2t}dp \wedge dr$$

$$= \frac{1}{pr}dp \wedge dr$$

$$= \omega.$$

The matrix associated with ω is

$$\begin{pmatrix} 0 & \frac{1}{pr} \\ -\frac{1}{pr} & 0 \end{pmatrix}.$$

X_H for this Γ_t can be found by

$$X_H(p,r) = \frac{d}{dt}\Gamma_t(p,r)|_{t=0} = \begin{pmatrix} p \\ r \end{pmatrix}.$$

We can now find dH by Equation 4.4, the symplectic geometer's version of Hamilton's equation.

$$dH = -\iota_{X_H}\omega$$

$$= -\begin{pmatrix} p & r \end{pmatrix}\begin{pmatrix} 0 & \frac{1}{pr} \\ -\frac{1}{pr} & 0 \end{pmatrix}$$

$$= -\begin{pmatrix} -\frac{1}{p} \\ \frac{1}{r} \end{pmatrix}$$

$$= \begin{pmatrix} \frac{1}{p} \\ -\frac{1}{r} \end{pmatrix}.$$

This tells us that $\frac{\partial H}{\partial p} = -\frac{1}{p}$ and $\frac{\partial H}{\partial r} = \frac{1}{r}$. The function H satisfying these relations is

$$H = \ln(\frac{r}{p}) + C$$

where C is a constant.

Exercise 76 *Let T be a positive real number. Find all Hamiltonian functions on the cylinder whose associated flows (using the area form) rotate the cylinder steadily with period T.*

Solution

There are only two flows that rotate the cylinder steadily with period T, namely

$$\begin{pmatrix} p \\ \theta \end{pmatrix} \mapsto \begin{pmatrix} p \\ \theta \pm \frac{2\pi}{T}t \end{pmatrix},$$

where one flow has a plus sign and one flow a minus sign. Using the plus/minus sign ($\pm$) allows us to do the two calculations for the two flows simultaneously. By differentiating the expression above we find that the velocity vector of the flow is $X = \pm\frac{2\pi}{T}\frac{\partial}{\partial\theta}$. So we need to find all functions H such that

$$dH = -\iota_X\omega = \iota_{\frac{2\pi}{T}\frac{\partial}{\partial\theta}} R\,dp \wedge dr = \pm\frac{2\pi R}{T}dp.$$

Integrating both sides we find that the function H must be of the form

$$H = \pm\frac{2\pi R}{T}p + C$$

for some constant C.

5   Symmetries are Lie Group Actions

Exercise 79 *For any natural number n the set of invertible $n \times n$ matrices with rational entries is a group, denoted $GL(n, \mathbb{Q})$. Show (for every n) that $GL(n, \mathbb{Q})$ is a group but not a matrix Lie group.*

Solution

To show show that the set $GL(n, \mathbb{Q})$ is a group, we must verify that the set is closed under matrix multiplication, and that it satisfies the axioms given in Definition 16.

To show $GL(n, \mathbb{Q})$ is closed under matrix multiplication, consider matrices g and h in $GL(n, \mathbb{Q})$. Each entry of gh is obtained from sums and products of entries in g and h. Since $\mathbb{Q}$ is closed under multiplication and addition, gh must have rational entries. Also, since $\det(gh) = \det g \det h$, gh must be invertible. So $gh \in GL(n, \mathbb{Q})$, meaning $GL(n, \mathbb{Q})$ is closed under matrix multiplication.

Axiom 1 of Definition 15 is verified by the associativity of matrix multiplication. Axiom 2 is verified by letting I be the $n \times n$ identity matrix. Axiom 3 is verified by letting g^{-1} be the inverse matrix of g. Note that the entries of g^{-1} are rational functions of the entries of g; hence g^{-1} will have rational coefficients if g does. Also, g^{-1} is invertible since $(g^{-1})^{-1} = g$. Hence, $g^{-1} \in GL(n, \mathbb{Q})$. This proves that $GL(n, \mathbb{Q})$ is a group.

However, $GL(n, \mathbb{Q})$ is not a matrix Lie group. To see this, look at the following sequence in $GL(2, \mathbb{Q})$:

$$\begin{pmatrix} 3 & 0 \\ 0 & 1 \end{pmatrix}, \begin{pmatrix} 3.1 & 0 \\ 0 & 1 \end{pmatrix}, \begin{pmatrix} 3.14 & 0 \\ 0 & 1 \end{pmatrix}, \begin{pmatrix} 3.141 & 0 \\ 0 & 1 \end{pmatrix}, \dots$$

where the nth matrix has a_{11} equal to the $(n-1)$th decimal approximation of π. This sequence is contained in $GL(2, \mathbb{Q})$, but its limit,

$$\begin{pmatrix} \pi & 0 \\ 0 & 1 \end{pmatrix},$$

is not. So $GL(2, \mathbb{Q})$ is not closed under taking limits within the set of invertible matrices, hence it is not a matrix Lie group. Similar examples show that $GL(n, \mathbb{Q})$ is not a matrix Lie group for any n.

Exercise 80 *Show that every matrix Lie group is a group.*

Solution

Let G be a matrix Lie group. By the definition of matrix Lie groups, G is closed under matrix mulitiplication. To verify that G is a group, we must show it satisfies the three axioms of Definition 15. Axiom 1 is verified by the associativity of matrix multiplication. Axiom 2 is verified by letting I be the $n \times n$ identity matrix. Axiom 3 is verified by letting g^{-1} be the inverse matrix of g. Note that by the definition of a matrix Lie group, we know $g^{-1} \in G$. This proves that G is a group.

Exercise 82 *Show that $(\mathbb{R}, \times)$ is neither a group nor a matrix Lie group.*

Solution

$(\mathbb{R}, \times)$ is not a group because the element $0 \in \mathbb{R}$ has no inverse. $(\mathbb{R}, \times)$ is not a matrix Lie group since we proved in Exercise 80 that every matrix Lie group must also be a group. Another reason $(\mathbb{R}, \times)$ is not a matrix Lie group is that there is no inverse matrix of 0, so $(\mathbb{R}, \times)$ is not closed under matrix inversion.

Exercise 84 *Show that both S^1 and $SO(2)$ are manifolds and that the function above and its inverse are both infinitely differentiable.*

Solution

From Chapter 3 we know that S^1 is a manifold via coordinate patches γ_1 : $(0, 2\pi) \to \mathbb{C}, t_1 \mapsto \cos t_1 + i \sin t_1$ and $\gamma_2 : (-\pi, \pi) \to \mathbb{C}, t_2 \mapsto \cos t_2 + i \sin t_2$. Similarly, we can show that $SO(2)$ is a manifold via coordinate patches

$$\beta_1 : (0, 2\pi) \to 2 \times 2 \text{ matrices}$$
$$\theta_1 \mapsto \begin{pmatrix} \cos \theta_1 & -\sin \theta_1 \\ \sin \theta_1 & \cos \theta_1 \end{pmatrix}$$

and

$$\beta_2 : (-\pi, \pi) \to 2 \times 2 \text{ matrices}$$
$$\theta_2 \mapsto \begin{pmatrix} \cos \theta_2 & -\sin \theta_2 \\ \sin \theta_2 & \cos \theta_2 \end{pmatrix}$$

In the $\gamma_1 \beta_1$-patch, the function in question is $(0, 2\pi) \to (0, 2\pi), t_1 \mapsto t_1$, which is infinitely differentiable. In the $\gamma_1 \beta_2$-patch the function is

$$(-\pi, 0) \cup (0, \pi) \to (0, 2\pi)$$
$$t \mapsto \begin{cases} t_1 + 2\pi, & t_1 < 0 \\ t_1, & t_1 > 0. \end{cases}$$

This function is infinitely differentiable at every point of its domain. Note that the problematic point 0 does not lie in the domain! The analysis of the other two coordinate patches is similar. We conclude that the given function is infinitely differentiable.

A similar argument applies to the inverse function and shows it infinitely differentiable.

Exercise 88 *Show that the Lie group of matrices of the form*

$$\begin{pmatrix} \cos(\sqrt{\frac{k}{m}}t) & \frac{1}{\sqrt{km}} \sin(\sqrt{\frac{k}{m}}t) \\ -\sqrt{km} \sin(\sqrt{\frac{k}{m}}t) & \cos(\sqrt{\frac{k}{m}}t) \end{pmatrix}.$$

is isomorphic to the circle group S^1.

Solution

We have shown that S^1 is isomorphic to the Lie group $SO(2)$ of rotations of 2-space. We will now show that $SO(2)$ is isomorphic to the group of matrices of the form

$$\begin{pmatrix} \cos(\sqrt{\frac{k}{m}}t) & \frac{1}{\sqrt{km}}\sin(\sqrt{\frac{k}{m}}t) \\ -\sqrt{km}\sin(\sqrt{\frac{k}{m}}t) & \cos(\sqrt{\frac{k}{m}}t) \end{pmatrix},$$

which we will denote by G. Since isomorphisms are transitive, this will show that S^1 is isomorphic to G.

Consider the invertible matrix

$$M = \begin{pmatrix} 0 & -\frac{1}{\sqrt{km}} \\ 1 & 0 \end{pmatrix}.$$

We define our isomorphism to be

$$f : SO(2) \to G$$
$$f : A \mapsto MAM^{-1}.$$

The reader may verify that this correspondence takes the matrix

$$\begin{pmatrix} \cos t & -\sin t \\ \sin t & \cos t \end{pmatrix}.$$

to the matrix

$$\begin{pmatrix} \cos(\sqrt{\frac{k}{m}}t) & \frac{1}{\sqrt{km}}\sin(\sqrt{\frac{k}{m}}t) \\ -\sqrt{km}\sin(\sqrt{\frac{k}{m}}t) & \cos(\sqrt{\frac{k}{m}}t) \end{pmatrix}$$

and is hence one-to-one and onto. Note that

$$f(A)f(B) = (MAM^{-1})(MBM^{-1}) = MABM^{-1} = f(AB),$$

which indicates that f preserves group multiplication. A similar argument shows that f preserves inversion. Therefore, f is an isomorphism.

Exercise 89 *Let G denote a matrix Lie group of $n \times n$ matrices. Define $S_g(v) := vg^T$ for any $g \in G$ and any row vector v in $\mathbb{R}^n$. Show that S_g defines an action of G on $(\mathbb{R}^n)^*$.*

Solution

Note that the correspondence $S_g(v) = vg^T$ takes vectors in $(\mathbb{R}^n)^*$ to vectors in $(\mathbb{R}^n)^*$. To show that this is a group action, we must show that this correspndence satisfies the two axioms of definition 18. Axiom 1 is easily verified, since $S_I(v) = vI^T = vI = v$. To verify axiom 2, consider elements $g, h \in G$. We have

$$S_g(S_h(v)) = S_g(vh^T) = vh^T g^T = v(gh)^T = S_{gh}(v).$$

Hence S_g defines an action of G on $(\mathbb{R}^n)^*$.

Exercise 92 *Given any vectors* **v**, **w** *in* $\mathbb{R}^3$ *with* $|\mathbf{v}| = |\mathbf{w}|$, *give a formula for a matrix g in* $SO(3)$ *such that* $g\mathbf{v} = \mathbf{w}$. *(Hint: consider an orthonormal basis of* $\mathbb{R}^3$ *containing* **v**$/|\mathbf{v}|$ *and another orthonormal basis containing* **w**$/|\mathbf{w}|$*).*

Solution

Let us assume, without loss of generality, that **v** and **w** are unit vectors. Let $B_1 = \mathbf{v}, \mathbf{a}, \mathbf{b}$ be one right-handed orthonormal basis for $\mathbb{R}^3$ and $B_2 = \mathbf{w}, \mathbf{c}, \mathbf{d}$ be another. Then

$$M_1 = \begin{pmatrix} \mathbf{v} & | & \mathbf{a} & | & \mathbf{b} \end{pmatrix} \text{ and } M_2 = \begin{pmatrix} \mathbf{w} & | & \mathbf{c} & | & \mathbf{d} \end{pmatrix}$$

are the matrices for the change of bases from B_1 to the standard basis and B_2 to the standard basis, respectively. So the matrix $M_2 M_1^{-1}$ will take **v**, expressed in the standard basis, to **w**, expressed in the standard basis. Note that since B_1 and B_2 are right-handed orthonormal bases, M_1 and M_2 are in $SO(3)$. Hence $M_2^{-1}M_1 \in SO(3)$, which completes this problem.

Exercise 94 *Consider the action of the group* $(\mathbb{Z}, +)$ *of integers acting on the real line* $\mathbb{R}$ *by translation: for each* $g \in \mathbb{Z}$ *and each* $r \in \mathbb{R}$ *we define* $S_g(r) := r + g$. *Show that* $\mathbb{R}/\mathbb{Z}$ *is isomorphic as a group to the circle group* S^1.

Solution

First we need to define a group structure on $\mathbb{R}/\mathbb{Z}$. We will use the structure that $\mathbb{R}/\mathbb{Z}$ inherits from the group $(\mathbb{R}, +)$. So for $a, b \in \mathbb{R}$, we define

$$\mathcal{O}_a + \mathcal{O}_b := \mathcal{O}_{a+b}.$$

We leave it to the reader to verify that this operation is well-defined (i.e., if $\mathcal{O}_a = \mathcal{O}_b$ and $\mathcal{O}_c = \mathcal{O}_d$ then $\mathcal{O}_{a+c} = \mathcal{O}_{b+d}$) and that $\mathbb{R}/\mathbb{Z}$ is a group under this operation. The group $\mathbb{R}/\mathbb{Z}$ is isomorphic to S^1 under the correspondence

$$f : \mathbb{R}/\mathbb{Z} \to S^1$$
$$f : \mathcal{O}_a \mapsto e^{2\pi a}.$$

We must first show that f is well-defined. Given $a, b \in \mathbb{R}$ such that $\mathcal{O}_a = \mathcal{O}_b$, it must be true that $a \in \mathcal{O}_b$, hence $a = S_g(b) = b + g$ for some $g \in \mathbb{Z}$. But then

$$\begin{aligned} f(a) &= e^{2\pi a} \\ &= e^{2\pi(b+g)} \\ &= e^{2\pi b} e^{2\pi g} \\ &= e^{2\pi b} \cdot 1 \\ &= f(b). \end{aligned}$$

So f is well-defined. To check that f is one-to-one, consider $a, b \in \mathbb{R}$ such that $e^{2\pi a} = e^{2\pi b}$. Then $e^{2\pi(a-b)} = 1$, so $a - b$ must be an integer. If we call this integer g, we get that $a = S_g(b)$ and $b = S_{-g}(a)$, which proves that $\mathcal{O}_a = \mathcal{O}_b$. To verify that f is onto, just note that for every $x \in S^1$, $x = e^{2\pi a} = f(\mathcal{O}_a)$ for some $a \in \mathbb{R}$. Finally, we must show that f preserves the group operation. Given $a, b \in \mathbb{R}$, we have

$$f(\mathcal{O}_a + \mathcal{O}_b) = f(\mathcal{O}_{a+b}) = e^{2\pi(a+b)} = e^{2\pi a} e^{2\pi b} = f(a) f(b).$$

Therefore, f preserves the group operation and is an isomorphism.

6   Infinitesimal Symmetries are Lie Algebras

Exercise 98 *Show that for any infinitesimal matrix ξ (not necessarily antisymmetric), we have $\det(I+\xi) = 1$ if and only if $\operatorname{tr} \xi = 0$. Conclude that the matrix Lie algebra of the matrix Lie group $SL(n, \mathbb{R})$ is*

$$sl(n, \mathbb{R}) := \{\xi : \xi \text{ is an } n \times n \text{ real matrix and } \operatorname{tr} \xi = 0\} .$$

Solution

Note that for any infinitesimal matrix ξ, $\det(I + \xi)$ is equal to the product of the diagonal elements of $I + \xi$. This is true because every entry of $I + \xi$ off the diagonal is either infinitesimal or zero, and every term in $\det(I + \xi)$ besides the product of the diagonal entries will contain at least two of these terms. Therefore, if we let ξ_{ij} denote the entries of ξ, we get that

$$\det(I + \xi) = \prod_{i=1}^{n}(1 + \xi_{ii}),$$

where n is the dimension of xi. Expanding the left side of the above equation and removing all terms of order 2 or higher, we get

$$\det(I + \xi) = 1 + \sum_{i=1}^{n}\xi_{ii}.$$

Thus, $\det(I + \xi) = 1$ if and only if $\text{tr}(\xi) = 0$. Since $SL(n, \mathbb{R})$ is defined as the set of $n \times n$ matrices with determinant 1, we can conclude that $sl(n, \mathbb{R})$ is the set of $n \times n$ matrices with zero trace.

Exercise 99 *Show that the matrix Lie algebra of the matrix Lie group $GL(n, \mathbb{R})$ is*

$$gl(n, \mathbb{R}) := \{\xi : \xi \text{ is an } n \times n \text{ real matrix.}\}$$

Solution

$GL(n, \mathbb{R})$ is the set of all $n \times n$ matrices with nonzero determinant. So to find $gl(n, \mathbb{R})$ we will first find the set of all infinitesimal matrices ξ such that $\det(I + \xi) \neq 0$. In the solution to the previous problem we proved that

$$\det(I + \xi) = 1 + \sum_{i=1}^{n}\xi_{ii}.$$

However, the infinitesimals are "small enough" so that the absolute value of their sum will be "small." So any infinitesimal matrix ξ will satisfy $\det(I+\xi) \neq 0$, and we conclude that $gl(n, \mathbb{R})$ is the set of all $n \times n$ matrices.

Exercise 100 *Prove the following facts about the antisymmetric 3×3 matrix $\tilde{\xi}$ and the column vector $\tilde{\xi}$:*

1. *For any vector $\mathbf{r}$ we have $\xi\mathbf{r} = \tilde{\xi} \times \mathbf{r}$.*

2. *The vector $\tilde{\xi}$ is an eigenvector for ξ with eigenvalue 0.*

3. *If a vector $\mathbf{v}$ is perpendicular to $\tilde{\xi}$, then the vector $\xi\mathbf{v}$ is also perpendicular to $\tilde{\xi}$. In other words, the plane of vectors perpendicular to $\tilde{\xi}$ is an invariant space for the matrix ξ. Call this plane $\tilde{\xi}^{T}$.*

4. *If we restrict the linear transformation ξ to the plane $\tilde{\xi}^T$, we get an infinitesimal rotation of that plane, and any infinitesimal rotation of any plane comes from some antisymmetric matrix.*

Solution

The first two parts of this problem are straightforward matrix computation.

3. It is in fact true that for any vector $\mathbf{v}$, $\xi\mathbf{v}$ is perpendicular to $\tilde{\xi}$. This is true because

$$(\xi\mathbf{v})^T\tilde{\xi} = \mathbf{v}^T\xi^T\tilde{\xi} = -\mathbf{v}^T\xi\tilde{\xi} = 0.$$

4. The group of rotations of the plane $\tilde{\xi}^T$ is the subgroup of $SO(3)$ which leaves $\pm\tilde{\xi}$ fixed, that is, the group of all $g \in SO(3)$ such that $g\tilde{\xi} = \tilde{\xi}$. Hence, the conditions for a matrix A to be an infinitesimal rotation of $\tilde{\xi}^T$ are (1) $A \in so(3)$ and (2) $(I + A)\tilde{\xi} = \tilde{\xi}$. We already know that $\xi \in so(3)$, and it is easy to verify that

$$(I + \xi)\tilde{\xi} = \tilde{\xi} + \xi\tilde{\xi} = \tilde{\xi} + 0 = \tilde{\xi}.$$

Hence ξ is an infinitesimal rotation of $\tilde{\xi}$.

To answer the second part of this question, simply observe that any rotation of any plane in three-space extends to a rotation of three-space. This fact should be fairly intuitive, but if the reader would like an explicit formula for this extension, here it is: Suppose M is a rotation of the plane p which lies in $\mathbb{R}^3$. Any vector in $\mathbb{R}^3$ can be written as $\mathbf{v} + \mathbf{w}$ where $\mathbf{v}$ lies in p and $\mathbf{w}$ is perpendicular to p. The rotation M of $\mathbf{v} + \mathbf{w}$ is defined to be $M\mathbf{v} + \mathbf{w}$.

By an analogous argument, any infinitesimal rotation of a plane in three-space can be extended to an infinitesimal rotation of three-space. We have already shown that infinitesimal rotations of three-space are represented by antisymmetric matrices, so infinitesimal rotations of planes in three-space must come from antisymmetric matrices.

Exercise 104 *Suppose that ω is the angular velocity of a rotating system. Suppose that ξ is a antisymmetric matrix with $\tilde{\xi} = \omega$. Show that ξ is the linear operator taking each position vector to its velocity. In other words, show that for all vectors $\mathbf{r} \in \mathbb{R}^3$ we have $\frac{d}{dt}\mathbf{r} = \xi\mathbf{r}$.*

Solution

For any unit vector $\mathbf{e}$ we have

$$\frac{d}{dt}\mathbf{e} = \omega \times \mathbf{e} = \xi\mathbf{e},$$

Let $\mathbf{r}$ denote an arbitrary vector rotating with angular velocity ω. Since rotations preserve lengths, there is a nonnegative scalar c, constant in time, and a unit vector $\mathbf{e}$, depending on time, such that $\mathbf{r} = c\mathbf{e}$ for all times. It is intuitively clear that $\mathbf{e}$ rotates with angular velocity ω, so $\frac{d}{dt}\mathbf{e} = \xi\mathbf{e}$. Hence

$$\frac{d}{dt}\mathbf{r} = \frac{d}{dt}(c\mathbf{e}) = c\frac{d}{dt}\mathbf{e} = c\xi\mathbf{e} = \xi c\mathbf{e} = \xi\mathbf{r}.$$

Exercise 106 *For each element ξ of $\mathbb{R}^3$, construct a parametrized path g through the identity in the matrix Lie group G so that*

$$g'(0) = \begin{pmatrix} 0 & 0 & 0 & \xi_x \\ 0 & 0 & 0 & \xi_y \\ 0 & 0 & 0 & \xi_z \\ 0 & 0 & 0 & 0 \end{pmatrix}.$$

Solution

Given $\xi \in \mathbb{R}^3$, the path

$$g(s) = \begin{pmatrix} 1 & 0 & 0 & \xi_x s \\ 0 & 1 & 0 & \xi_y s \\ 0 & 0 & 1 & \xi_z s \\ 0 & 0 & 0 & 1 \end{pmatrix}$$

in G has $g(0) = I$ and

$$g'(0) = \begin{pmatrix} 0 & 0 & 0 & \xi_x \\ 0 & 0 & 0 & \xi_y \\ 0 & 0 & 0 & \xi_z \\ 0 & 0 & 0 & 0 \end{pmatrix}.$$

Exercise 107 *Given a matrix Lie group G and an element ξ of its Lie algebra, the one-parameter subgroup generated by ξ is*

$$G_\xi := \left\{ e^{s\xi} : s \in \mathbb{R} \right\}.$$

For which $\xi \in so(3)$ is $G_\xi \cong S^1$?

Solution

By Exercise 92 there is a matrix $g \in SO(3)$ such that $g\begin{pmatrix} 0 \\ 0 \\ 1 \end{pmatrix}$ equals the

vector $\tilde{\xi}$ defined in Exercise 100. Then $g^{-1}\xi g$ is an antisymmetric 3×3 matrix, and $g^{-1}\xi g \begin{pmatrix} 0 \\ 0 \\ 1 \end{pmatrix} = \mathbf{0}$. So $g^{-1}\xi g$ is an infinitesimal rotation around the z-axis

and, as we already know from the text, the group

$$\{e^{sg^{-1}\xi g} : s \in \mathbb{R}\}$$

is a circle group. But it is not hard to check (using the series expansion formula for matrix exponentiation) that $e^{sg^{-1}\xi g} = g^{-1}e^{s\xi}g$, and hence this circle group is isomorphic to G_ξ. Hence G_ξ is a circle group.

Exercise 110 *Check that if $A \in so(3)$ and $B \in so(3)$, then $[A, B] \in so(3)$.*

Solution

$so(3)$ is the set of 3×3 antisymmetric matrices. So for $A, B \in so(3)$,

$$
\begin{aligned}
[A, B]^T &= (AB - BA)^T = (AB)^T - (BA)^T \\
&= A^T B^T - B^T A^T = AB - BA = -[A, B].
\end{aligned}
$$

This proves that $[A, B] \in so(3)$.

7   Conserved Quantities are Momentum Maps

Exercise 112 *Let V be a vector space of matrices all of the same dimensions. Show that for each matrix A of the same dimensions as the elements of V the function $V \to \mathbb{R}$, $B \mapsto \operatorname{tr}(A^T B)$ is a linear functional.*

Solution

The identities $tr(M + N) = tr(M) + tr(N)$ and $tr(aM) = a\,tr(M)$ are easily verified for scalars a and matrices M and N of the same dimensions. Using these identities, we can see that, for any elements B and C of the vector space V and for any scalar a, we have

$$tr(A^T(B + C)) = tr(A^T B + A^T C) = tr(A^T B) + tr(A^T C)$$

and

$$tr(A^T(aB)) = tr(aA^T B) = a\,tr(A^T B).$$

Hence the function in question is a linear functional.

Exercise 113 *Show that the translation action*

$$S_{\mathbf{g}} : (\mathbf{r}_1, \mathbf{r}_2, \mathbf{p}_1, \mathbf{p}_2) \mapsto (\mathbf{r}_1 + \mathbf{g}, \mathbf{r}_2 + \mathbf{g}, \mathbf{p}_1, \mathbf{p}_2)$$

does not preserve the one-form $\mathbf{r}_1 d\mathbf{p}_1 + \mathbf{r}_2 d\mathbf{p}_2$.

Solution

If we pull the one-form $\mathbf{r_1}dp_1 + \mathbf{r_2}dp_2$ back under $S_\mathbf{g}$ we get

$$
\begin{aligned}
S_\mathbf{g}^*(\mathbf{r_1}dp_1 + \mathbf{r_2}dp_2) &= (\mathbf{r_1} + \mathbf{g})dp_1 + (\mathbf{r_2} + \mathbf{g})dp_2 \\
&= \mathbf{r_1}dp_1 + \mathbf{r_2}dp_2 + \mathbf{g}(dp_1 + dp_2),
\end{aligned}
$$

which is not equal to $\mathbf{r_1}dp_1 + \mathbf{r_2}dp_2$ because neither $\mathbf{g}$ nor $dp_1 + dp_2$ is equal to zero.

Exercise 114 *Let T^2 be the two-dimensional torus, $T^2 = S^1 \times S^1$. Parametrize the first circle by the angle θ and the second by the angle ψ. Let $(\mathbb{R}, +)$ act on T^2 by*

$$
S_t : (\theta, \psi) \mapsto (\theta, \psi + t).
$$

Describe this action geometrically. Extra credit: show that this action has no momentum map.

Solution

If we say that ψ is the angle going around the outside of the torus and θ is the angle going through the middle, then the action S_t is a rotation of the torus around the axis going through the center of its hole.

Extra credit: We will show that there is no Hamiltonian H on the torus inducing the vector field defined by S_t, which will prove that there is no momentum map inducing this vector field. These results will be independent of the symplectic form ω on the torus.

Assume there is such an H. Our geometric intuition tells us (correctly) that for any ξ in the Lie algebra of $(\mathbb{R}, +)$ (which is just $(\mathbb{R}, +)$ itself), the vector field on the torus will be a field of vectors pointing around the outside of the torus in the direction of increasing ψ. The number ξ determines only the magnitude of these vectors.

Pick any point on the torus. The equation

$$
i_{X_H}\omega = -dH
$$

must be true at this point. We know that $\omega(X_H, \frac{\partial}{\partial\psi}) = 0$ since X_H is in the same direction as $\frac{\partial}{\partial\psi}$. The nondegeneracy of ω therefore implies that $\omega(X_H, \frac{\partial}{\partial\theta})$ is nonzero since $\frac{\partial}{\partial\psi}$ and $\frac{\partial}{\partial\theta}$ form a basis for local coordinates on the torus. We have therefore shown that $dH\frac{\partial}{\partial\theta}$ is never zero.

Now consider a circle going through the center of the torus formed by holding ψ constant and letting θ vary from 0 to 2π. Let $h(\theta)$ denote the value of H at a particular value of θ on this circle. Note that $h'(\theta) = dH \frac{\partial}{\partial \theta}$. Obviously, $h(0) = h(2\pi)$, since $\theta = 0$ and $\theta = 2\pi$ represent the same point on this circle. Therefore, by the mean value theorem, $h'(\theta)$ must be equal to zero for some $\theta, 0 \le \theta \le 2\pi$. But this contradicts our assertion that $dH \frac{\partial}{\partial \theta}$ is never zero. Hence no such H exists.

8   Reduction and the Two-Body Problem

Exercise 115 *Use the implicit function theorem to show that for any nonzero $\tilde{\mathbf{L}} \in \mathbb{R}^3$ the set*

$$S_{\tilde{\mathbf{L}}}\{(\mathbf{r}, \mathbf{p}) \in \mathbb{R}^6 : \mathbf{r} \times \mathbf{p}^T = \tilde{\mathbf{L}}\}$$

can be parametrized locally differentiably near any point by three parameters. In other words, the set is a three-dimensional manifold. What can you say about the set S_0? Is it a manifold?

Solution

We will apply the implicit function theorem to the function $f : \mathbb{R}^6 \to \mathbb{R}^3$ defined by

$$f(\mathbf{r}, \mathbf{p}^T) := \mathbf{r} \times \mathbf{p}^T.$$

Note that

$$Df_{(\mathbf{r},\mathbf{p}^T)} = \begin{pmatrix} 0 & -r_z & r_y & 0 & -p_z & p_y \\ r_z & 0 & -r_x & p_z & 0 & -p_x \\ -r_y & r_z & 0 & -p^y & p_z & 0 \end{pmatrix}.$$

Because $\tilde{\mathbf{L}} \neq 0$ we know that $\mathbf{r}$ and $\mathbf{p}^T$ are not parallel, $\mathbf{r} \neq 0$ and $\mathbf{p}^T \neq 0$. Hence the matrix $Df_{(\mathbf{r},\mathbf{p}^T)}$ has rank three (and hence corank three) for any $(\mathbf{r}, \mathbf{p}^T)$ such that $f(\mathbf{r}, \mathbf{p}^T) = \tilde{\mathbf{L}}$. So, by the inverse function theorem, the surface $S_{\tilde{\mathbf{L}}}$ can be parametrized locally differentiably by three parameters near any point.

The set S_0 is not a manifold. Intuitively we can argue that if S_0 were a manifold then it would have to be a 4-manifold, since near any point $(\mathbf{r}, \mathbf{p}^T)$ with $\mathbf{r} \neq 0$ we can use the position of $\mathbf{r}$ and the ratio $\frac{|\mathbf{p}|}{|\mathbf{r}|}$ to define a coordinate patch from a subset of $\mathbb{R}^4$ to S_0. However, there would be at least six linearly independent vectors in the tangent space to S_0 at $(0, 0)$, namely, $(\mathbf{r}, 0)$ for any $\mathbf{r} \in \mathbb{R}^3$

and $(0, \mathbf{p}^T)$ for any $\mathbf{p}^T \in \mathbb{R}^3$. But a 4-manifold can have at most four linearly independent vectors in a single tangent space. Hence S_0 is not a manifold.

Exercise 116 *Check (using the definitions in the text) that $\frac{\mathbf{pr}}{|\mathbf{r}|} = \sigma$, that $\mathbf{p}\tilde{\mathbf{L}}=0$ and that $\frac{\mathbf{p}(\mathbf{r}\times\tilde{\mathbf{L}})}{|\mathbf{r}\times\tilde{\mathbf{L}}|} = -\frac{|\tilde{\mathbf{L}}|}{\rho}$.*

Solution

$\frac{\mathbf{pr}}{|\mathbf{r}|} = \sigma$ by the definition of σ. We know that $\mathbf{r} \times \mathbf{p}^T = \tilde{\mathbf{L}}$, which indicates that $\tilde{\mathbf{L}}$ is perpendicular to $\mathbf{p}$, and hence $\mathbf{p}^T\tilde{\mathbf{L}} = 0$. As for the third equation, note that $\mathbf{p}(\mathbf{r} \times \tilde{\mathbf{L}})$ is the triple scalar product of $\mathbf{p}^T$, $\mathbf{r}$, and $\tilde{\mathbf{L}}$, and is hence equal to $\tilde{\mathbf{L}}^T(\mathbf{p}^T \times \mathbf{r})$. So we have

$$\frac{\mathbf{p}(\mathbf{r} \times \tilde{\mathbf{L}})}{|\mathbf{r} \times \tilde{\mathbf{L}}|} = \frac{\tilde{\mathbf{L}}^T(\mathbf{p}^T \times \mathbf{r})}{|\mathbf{r} \times \tilde{\mathbf{L}}|}$$
$$= \frac{\tilde{\mathbf{L}}^T(-\tilde{\mathbf{L}})}{|\mathbf{r} \times \tilde{\mathbf{L}}|}$$
$$= -\frac{|\tilde{\mathbf{L}}|^2}{|\mathbf{r} \times \tilde{\mathbf{L}}|}.$$

However, $\mathbf{r}$ and $\tilde{\mathbf{L}}$ are orthogonal, and hence $|\mathbf{r} \times \tilde{\mathbf{L}}| = |\mathbf{r}||\tilde{\mathbf{L}}|$. Therefore, by the definition of ρ,

$$\frac{\mathbf{p}(\mathbf{r} \times \tilde{\mathbf{L}})}{|\mathbf{r} \times \tilde{\mathbf{L}}|} = -\frac{|\tilde{\mathbf{L}}|}{|\mathbf{r}|} = -\frac{|\tilde{\mathbf{L}}|}{\rho}.$$

Exercise 120 *Find the orbit of $SO(3)$ through $(0, 0)$. Find the orbit of $SO(3)$ through an arbitrary nonzero point in $\Phi_1^{-1}(0)$.*

Solution

The action of $SO(3)$ on M_1 is described by

$$S_g : (\mathbf{r}, \mathbf{p}) \mapsto (g\mathbf{r}, \mathbf{p}g^T).$$

So for any $g \in SO(3)$, we will have $S_g(0, 0) = (g0, 0g^T) = (0, 0)$. The orbit of $SO(3)$ through $(0, 0)$ will therefore be the point $(0, 0)$.

An arbitrary nonzero point in $\Phi_1^{-1}(0)$ will be a point $(\mathbf{r}, \mathbf{p})$ such that $\mathbf{r} \times \mathbf{p}^T = 0$, in other words, a point $(\mathbf{r}, \mathbf{p})$ such that $\mathbf{p}^T$ is parallel to $\mathbf{r}$. Let us find the orbit through such a point.

First we consider the case $\mathbf{r} \neq 0$. For any $g \in SO(3)$ we have $|g\mathbf{r}| = |\mathbf{r}|$ and $|(\mathbf{p}g^T)^T| = |\mathbf{p}^T|$. So every point (r_1, p_1) on the orbit satisfies $|r_1| = |\mathbf{r}|$

and $|p_1| = |p|$. Also, elements of $SO(3)$ preserve angles, so the angle (0 or π) between $\mathbf{r}$ and $\mathbf{p}^T$ must be the same as the angle between $\mathbf{r}_1$ and $\mathbf{p}_1^T$. Finally, by Exercise 6.3, for any $\mathbf{r}_1$ whose length equals the length of $\mathbf{r}$, there is an element $g \in SO(3)$ such that $g\mathbf{r} = r_1$. So the orbit through the point $(\mathbf{r}, \mathbf{p})$ is the sphere of position vectors of length $|r|$, with momentum vectors determined by their relationship to the position vectors.

In the case $\mathbf{r} = 0$ we have $\mathbf{p}^T \neq 0$ we again get a sphere: the sphere of pairs with position vector $\mathbf{0}$ and momentum vector of length $|p|$.

References

[AM]   Abraham, R. and J. Marsden, *Foundations of Mechanics, Second Edition*, Addison-Wesley, Reading, MA, 1978.

[Ap69]   Apostol, T. M., *Calculus, Volumes I and II, Second Edition*, Blaisdell, New York, 1969.

[Ap97]   Apostol, T. M., *Linear Algebra: A First Course, with Applications to Differential Equations*, John Wiley & Sons, Inc., New York, 1997.

[Ar89]   Arnold, V. I., *Mathematical Methods of Classical Mechanics, Second Edition*, Springer-Verlag, New York, 1989.

[Ar78]   Arnold, V. I., *Ordinary Differential Equations*, MIT Press, Cambridge, 1978.

[AG]   Arnold, V. I. and A. B. Givental, "Symplectic Geometry" in: *Encyclopaedia of Mathematical Sciences, Volume 4*, V. I. Arnold and S. P. Novikov, eds., Springer-Verlag, Berlin, 1990.

[Art]   Artin, M., *Algebra*, Prentice Hall, Upper Saddle River, New Jersey, 1991.

[At]   Atiyah, M. F., Convexity and commuting Hamiltonians, *Bulletin of the London Math. Soc.* **14** (1982), 1–15.

[Bai]   Baierlein, R, *Newtonian Dynamics*, McGraw-Hill, New York, 1983.

[BS]   Bamberg, P. and S. Sternberg, *A Course in Mathematics for Students of Physics, Volumes I and II*, Cambridge University Press, Cambridge, 1988.

[BD]   Boyce, W. E. and R. C. DiPrima, *Elementary Differential Equations and Boundary Value Problems, Fifth Edition*, Wiley, New York, 1992.

[Ba]   Bartle, R. G., *The Elements of Real Analysis, Second Edition*, Wiley, New York, 1976.

[BtD]   Bröcker, T. and T. tom Dieck, *Representations of Compact Lie Groups*, Springer-Verlag, New York, 1985.

[Br]   Bryant, R. L., "An introduction to Lie groups and symplectic geometry," in: *Geometry and Quantum Field Theory*, D. S. Freed and K. K. Uhlenbeck, eds., American Mathematical Society (IAS/Park City Mathematics Series), Providence, 1995.

[CB]   Cushman, R. H. and L. M. Bates, *Global Aspects of Classical Integrable Systems*, Birkhäuser Verlag, Basel, 1991.

[dC]   do Carmo, M. P., *Riemannian Geometry*, Birkhäuser, Boston, 1992.

[Fe]   Feynman, R. P., R. B. Leighton and M. Sands, *The Feynman Lectures on Physics*, Addison-Wesley, Reading, MA., 1964.

[Fr]   Frankel, Theodore, *The Geometry of Physics: An Introduction*, Cambridge University Press, Cambridge, England, 1997.

[Gi]   Ginzburg, V. L., Some remarks on symplectic actions of compact groups, *Math. Z.* **210** (1992), 625–640.

[Go]   Goldstein, H., *Classical Mechanics*, Addison-Wesley, Reading MA., 1950.

[Gr]   Greene, B., *The Elegant Universe: Superstrings, Hidden Dimensions, and the Quest for the Ultimate Theory*, Vintage Books (A Division of Random House, Inc.), New York, 1999.

[GS82]   Guillemin, V. and S. Sternberg, Convexity properties of the moment map, *Inventiones Math.* **67** (1982), 515–538.

[GS84] Guillemin, V. and S. Sternberg, *Symplectic Techniques in Physics*, Cambridge University Press, Cambridge, England, 1984.

[GS90] Guillemin, V. and S. Sternberg, *Variations on a Theme of Kepler*, American Mathematical Society Colloquium Publications, Vol. 42, American Mathematical Society, Providence, RI, 1990.

[HS]   Retrieved July 1, 2000 from the World Wide Web: http://sao-www.harvard.edu/cfa/ps/pressinfo/HaleBopp.html, Harvard-Smithsonian Center for Astrophysics, Cambridge, MA, 1997.

[I]   Iglesias, P.,"Les origines du calcul symplectique chez Lagrange," in: *Le journal de maths des elèves*, Volume 1 (1995), No. 3, pp. 153–161.

[K]   Kelley, J. L., *General Topology (Graduate Texts in Mathematics Series, Vol. 27)*, Springer-Verlag, New York, 1991.

[Ke1]   Kepler, J., *Astronomia nova, sev, Physica coelestis : tradita commentariis de motibvs stell, Martis, ex observationibus G.V. Tychonis Brahe, jussu & sumptibus Rvdolphi II., plurimum annorum pertinaci studio elaborata Prag* G. Voegelinus, Heidelberg, 1609.

[Ke2]   Kepler, J., *The Harmony of the World*, E. J. Aiton, A. M. Duncan and J. V. Field, trans., American Philosophical Society, Philadelphia, 1997.

[Koe]   Koestler, A., *The Watershed: A Biography of Johannes Kepler*, Doubleday & Co., Garden City, New York, 1960.

[Kos]   Kostant, B., "Quantization and Unitary Representations," in: *Lectures in Modern Analysis and Applications III*, LNM 170, Springer-Verlag, Berlin, 1970.

[HH]   Hubbard, J. H. and B. B. Hubbard, *Vector Calculus, Linear Algebra and Differential Forms: A Unified Approach*, Prentice Hall, Upper Saddle River, New Jersey, 1999.

[L]   Lay, D. C., *Linear Algebra and Its Applications (Second Edition)*, Addison-Wesley, Reading MA., 1997.

[LT]   Lerman, E. and S. Tolman, "Hamiltonian torus actions on symplectic orbifold and toric varieties", *Trans. Amer. Math. Soc.* 349 (1997), no. 10, 4201–4230.

[MH]   Marsden, J. E. and M. J. Hoffman, *Elementary Classical Analysis, Second Edition*, W. H. Freeman and Co., New York, 1993.

[MR]   Marsden, J. E. and T. S. Ratiu, *Introduction to Mechanics and Symmetry, Second Edition*, Springer-Verlag, New York, 1999.

[MTW] Marsden, J. E., A. J. Tromba and A. Weinstein, *Basic Multivariable Calculus*, Springer-Verlag, New York, 1993.

[MW]   Marsden, J. and A. Weinstein, Reduction of symplectic manifolds with symmetry, *Rep. Math. Phys.* **5** (1975), 121–130.

[MS]   McDuff, D. and D. Salamon, *Introduction to Symplectic Topology, Second Edition*, Clarendon Press, Oxford, 1998.

[Me]   Meyer, K. R., "Symmetries and Integrals in Mechanics", in: *Dynamical Systems*, M. M. Peixoto, ed., Academic Press, New York, 1973.

[Mo]   Moser, J., Regularization of Kepler's Problem and the Averaging Method on a Manifold, *Comm. Pure and Applied Math.*, **23** (1970), 609–636.

[Mu]   Munkres, J. R., *Analysis on Manifolds*, Addison-Wesley, Reading, MA., 1991.

[N]   Newton, I., *The Principia: Mathematical Principles of Natural Philosophy*, translated by I. B. Cohen and A. Whitman, Univ. of California Press, Los Angeles, 1999.

[O]   Olver, P. J., *Applications of Lie Groups to Differential Equations, Second Edition*, Springer-Verlag, New York, 1993.

[SSC]   Sanz-Serna, J. M., and M. P. Calvo, *Numerical Hamiltonian Problems*, Chapman & Hall, London, 1994.

[Sh]   Shenk, A., *Calculus and Analytic Geometry, Fourth Edition*, Scott, Foresman and Company, Glenview, Illinois, 1988.

[Si]   Simmons, G. F., *Differential Equations with Applications and Historical Notes, Second Edition*, McGraw Hill, Inc., New York, 1991.

[Sm]   Smale, S., Topology and Mechanics, I, *Inventiones Math.*, **10** (1970), 305–331.

[So]   Souriau, J.-M., *Structure des systèmes dynamiques*, Dunod, Paris, 1970, translated by C. H. Cushman-de Vries as *Structure of Dynamical Systems: A Symplectic View of Physics*, Birkhäuser, Boston, 1997.

[Sp]   Spivak, M., *Calculus, Third Edition*, Publish or Perish, Inc., Houston, Texas, 1994.

[St]   Strang, G., *Linear Algebra and Its Applications, Second Edition*, Academic Press, New York, 1980.

[Sz]   Szebehely, V. G., *Adventures in Celestial Mechanics: A First Course in the Theory of Orbits*, University of Texas Press, Austin, 1989.

[To]   Török, J. S., *Analytical Mechanics, with an Introduction to Dynamical Systems*, John Wiley & Sons, Inc., New York, 2000.

[Wa]   Warner, F. W., *Foundations of Differentiable Manifolds and Lie Groups*, Springer-Verlag, New York, 1983.

[Z]   Zee, A., *Fearful Symmetry: The Search for Beauty in Modern Physics*, Princeton Science Library, Princeton, NJ, 1986.

Index